Educational Measurement for Applied Researchers

Margaret Wu · Hak Ping Tam
Tsung-Hau Jen

Educational Measurement for Applied Researchers

Theory into Practice

Springer

Margaret Wu
National Taiwan Normal University
Taipei
Taiwan

and

Educational Measurement Solutions
Melbourne
Australia

Hak Ping Tam
Graduate Institute of Science Education
National Taiwan Normal University
Taipei
Taiwan

Tsung-Hau Jen
National Taiwan Normal University
Taipei
Taiwan

ISBN 978-981-10-3300-1 ISBN 978-981-10-3302-5 (eBook)
DOI 10.1007/978-981-10-3302-5

Library of Congress Control Number: 2016958489

Printed on acid-free paper

This Springer imprint is published by Springer Nature
The registered company is Springer Nature Singapore Pte Ltd.
The registered company address is: 152 Beach Road, #22-06/08 Gateway East, Singapore 189721, Singapore

Preface

This book aims at providing the key concepts of educational and psychological measurement for applied researchers. The authors of this book set themselves to a challenge of writing a book that covers some depths in measurement issues, but yet is not overly technical. Considerable thoughts have been put in to find ways of explaining complex statistical analyses to the layperson. In addition to making the underlying statistics accessible to non-mathematicians, the authors take a practical approach by including many lessons learned from real-life measurement projects. Nevertheless, the book is not a comprehensive text on measurement. For example, derivations of models and estimation methods are not dealt in detail in this book. Readers are referred to other texts for more technically advanced topics. This does not mean that a less technical approach to present measurement can only be at a superficial level. Quite the contrary, this book is written with considerable stimulation for deep thinking and vigorous discussions around many measurement topics. For those looking for recipes on how to carry out measurement, this book will not provide answers. In fact, we take the view that simple questions such as "how many respondents are needed for a test?" do not have straightforward answers. But we discuss the factors impacting on sample size and provide guidelines on how to work out appropriate sample sizes.

This book is suitable as a textbook for a first-year measurement course at the graduate level, since much of the materials for this book have been used by the authors in teaching educational measurement courses. It can be used by advanced undergraduate students who happened to be interested in this area. While the concepts presented in this book can be applied to psychological measurement more generally, the majority of the examples and contexts are in the field of education. Some prerequisites to using this book include basic statistical knowledge such as a grasp of the concepts of variance, correlation, hypothesis testing and introductory probability theory. In addition, this book is for practitioners and much of the content covered is to address questions we received along the years.

We would like to thank those who have made suggestions on earlier versions of the chapters. In particular, we would like to thank Tom Knapp and Matthias von Davier for going through several chapters in an earlier draft. Also, we would like to thank some students who had read several early chapters of the book. We benefit from their comments that help us to improve on the readability of some sections of the book. But, of course, any unclear spots or even possible errors are our own responsibility.

Taipei, Taiwan; Melbourne, Australia — Margaret Wu
Taipei, Taiwan — Hak Ping Tam
Taipei, Taiwan — Tsung-Hau Jen

Contents

Chapter 1
What Is Measurement?

Measurements in the Physical World

Most of us are familiar with measurement in the physical world, whether it is measuring today's maximum temperature, the height of a child or the dimensions of a house, where numbers are given to represent "quantities" of some kind, on some scales, to convey properties of some attributes that are of interest to us. For example, if yesterday's maximum temperature in London was 12 °C, one gets a sense of how cold (or warm) it was, without actually having to go to London in person to know about the weather there. If a house is situated 1.5 km from the nearest train station, one gets a sense of how far away that is, and how long it might take to walk to the train station. Measurement in the physical world is all around us, and there are well-established measuring instruments and scales that provide us with useful information about the world around us.

Measurements in the Psycho-social Science Context

Measurements in the psycho-social world are also abound, but perhaps less well established universally as temperature and distance measures. A doctor may provide a score for a measure of the level of depression. These scores may provide information to the patients, but the scores may not necessarily be meaningful to people who are not familiar with these measures. A teacher may provide a score of student achievement in mathematics. These may provide the students and parents with some information about progress in learning. But the scores will generally not provide much information beyond the classroom. The difficulty with measurement in the psycho-social world is that the attributes of interest are generally not directly visible to us as objects of the physical world are. It is only through observable indicator variables of the attributes that measurements can be made. For example, currently

M. Wu et al., *Educational Measurement for Applied Researchers*,
DOI 10.1007/978-981-10-3302-5_1

there is no machine that can directly measure depression. However, sleeplessness and eating disorders may be regarded as symptoms of depression. Through the observation of the symptoms of depression, one can then develop a measuring instrument and a scale of levels of depression. Similarly, to provide a measure of student academic achievement, one needs to find out what a student knows and can do academically. A test in a subject domain may provide us with some information about a student's academic achievement. One cannot "see" academic achievement as one sees the dimensions of a house. One can only measure academic achievement through indicator variables such as the performance on specific tasks by the students.

Psychometrics

From the above discussion, it can be seen that not only is the measurement of psycho-social attributes difficult, but often the attributes themselves are some "concepts" or "notions" which lack clear definitions. Typically, these psycho-social attributes need clarification before measurements can take place. For example, "academic achievement" needs to be defined before any measurement can be taken. In the following, psycho-social attributes to be measured are referred to as "latent traits" or "constructs". The science of measuring latent traits is referred to as psychometrics.

In general, psychometrics deals with the measurement of any "latent trait", and not just those in the psycho-social context. For example, the quality of wine has been an attribute of interest, and researchers have applied psychometric methodologies to establish a measurement scale for it. One can regard "the quality of wine" as a latent trait because it is not directly visible (therefore "latent"), and it is a concept that can have ratings from low to high (therefore "trait" to be measured) [see, for example, Thomson (2003)]. In general, psychometrics is about measuring latent traits where the attribute of interest is not directly visible so that the measurement is achieved through collecting information on indicator variables associated with the attribute. In addition, the attribute of interest to be measured varies in levels from low to high so that it is meaningful to provide "measures" of the attribute.

Before discussing the methods of measuring latent traits, it will be useful to examine some formal definitions of measurement and the associated properties of measurement. An understanding of the properties of measurement can help us build methodologies to achieve the best measurement in terms of the richness of information we can obtain from the measurement. For example, if the measures we obtain can only tell us whether a student's achievement is above or below average in his/her class, that's not a great deal of information. In contrast, if the measures can also inform us of the skills the student can perform, as well as how far ahead (or behind) he/she is in terms of yearly progression, then we have more information to act on to improve teaching and learning. The next section discusses properties of measurement with a view to identify the most desirable properties. In latter chapters of this book, methodologies to achieve good measurement properties are presented.

Formal Definitions of Psycho-social Measurement

Various formal definitions of psycho-social measurement can be found in the literature. The following are four different definitions of measurement. It is interesting to compare the scope of measurement covered by each definition.

- Measurement is a procedure for the assignment of numbers to specified properties of experimental units in such a way as to characterise and preserve specified relationships in the behavioural domain.
  Lord, F., & Novick, M. (1968) Statistical Theory of Mental Test Scores, p.17.
- Measurement is the assigning of numbers to individuals in a systematic way as a means of representing properties of the individuals.
  Allen, M.J. and Yen, W. M. (1979). Introduction to Measurement Theory, p 2.
- Measurement consists of rules for assigning numbers to objects in such a way as to represent quantities of attributes.
  Nunnally, J.C. & Bernstein, I.H. (1994) Psychometric Theory, p 1.
- Measurement begins with the idea of a variable or line along which objects can be positioned, and the intention to mark off this line in equal units so that distances between points on the line can be compared.
  Wright, B. D. & Masters, G. N. (1982). Rating Scale Analysis, p 1.

All four definitions relate measurement to assigning numbers to objects. The third and fourth definitions specifically bring in a notion of representing quantities, while the first and second state more generally the assignment of numbers in some well-defined ways. The fourth definition explicitly states that the quantity represented by the measurement is a continuous variable (i.e., on a real-number line), and not just a discrete rank-ordering of objects.

So it can be seen that the first and second definitions are broader and less specific than the third and the fourth. Measurements under the first and second definitions may not be very useful if the numbers are simply labels for objects since such measurements would not provide a great deal of information. The third and fourth definitions are restricted to "higher" levels of measurement in that the assignment of numbers can be called measurement only if the numbers represent quantities and possibly distances between objects' locations on a scale. This kind of measurement will provide us with more information in discriminating between objects in terms of the levels of the attribute the objects possess.

Levels of Measurement

More formally, there are definitions for four levels of measurement (nominal, ordinal, interval and ratio) in terms of the way numbers are assigned to objects and the inference that can be drawn from the numbers assigned. This idea was introduced by Stevens (1946). Each of these levels is discussed below.

Nominal

When numbers are assigned to objects simply as labels for the objects, the numbers are said to be nominal. For example, each player in a basketball team is assigned a number. The numbers do not mean anything other than for the identification of the players. Similarly, codes assigned for categorical variables such as gender (male = 1; female = 2) are all nominal. In this book, the assignment of nominal numbers to objects is not considered as measurement, because there is no notion of "more" or "less" in the representation of the numbers. The kind of measurement described in this book refers to methodologies for finding out "more" or "less" of some attribute of interest possessed by objects.

Ordinal

When numbers are assigned to objects to indicate ordering among the objects, the numbers are said to be ordinal. For example, in a car race, numbers are used to represent the order in which the cars finish the race. In a survey where respondents are asked to rate their responses, the numbers 0–3 are used to represent strongly disagree, disagree, agree and strongly agree. In this case, the numbers represent an ordering of the responses. Ordinal measurements are often used, such as for ranking students, or for ranking candidates in an election, or for arranging a list of objects in order of preferences. While ordering informs us of which objects have more (or less) of an attribute, ordering does not in general inform us of the quantities, or amount, of an attribute. If a line from low to high represents the quantity of an attribute, ordering of the objects does not position the objects on the line. Ordering only tells us the relative positions of the objects on the line.

Interval

When numbers are assigned to objects to indicate the differences in amount of an attribute the objects have, the numbers are said to represent interval measurement. For example, time on a clock provides an interval measure in that 7 o'clock is two hours away from 5 o'clock, and four hours from 3 o'clock. In this example, the numbers not only represent ordering, but also represent an "amount" of the attribute so that distances between the numbers are meaningful and can be compared. We will be able to compute differences between the quantities of two objects. While there may be a zero point on an interval measurement scale, the zero is typically arbitrarily defined and does not have a specific meaning. That is, there is generally no notion of a complete absence of an attribute. In the example about time on a clock, there is no meaningful zero point on the clock. Time on a clock may be better regarded as an interval scale. However, if we choose a particular time and regard it as a starting point to measure time span, the time measured can be regarded as forming a ratio measurement scale. In measuring abilities, we typically only have notions of very low ability, but not zero ability. For example, while a test score of zero indicates that a student is unable to answer any question correctly on a particular test, it does not necessarily mean that the student has zero ability in the latent trait being measured. Should an easier test be administered, the student may very well be able to answer some questions correctly.

Ratio

In contrast, measurements are at the ratio level when numbers represent interval measures with a meaningful zero, where zero typically denotes the absence of the attribute (no quantity of the attribute). For example, the height of people in cm is a ratio measurement. If Person A's height is 180 cm and Person B's height is 150 cm, we can say that Person A's height is 1.2 times of Person B's height. In this case, not only distances between numbers can be compared, the numbers can form ratios and the ratios are meaningful for comparison. This is possible because there is a zero on the scale indicating there is no existence of the attribute. Interestingly, while "time" is shown to have interval measurement property in the above example, "elapsed time" provides ratio measurements. For example, it takes 45 min to bake a large round cake in the oven, but it takes 15 min to bake small cupcakes. So the duration of baking a large cake is three times that of baking small cupcakes. Therefore, elapsed time provides ratio measurement in this instance. In general, a measurement may have different levels of measurement (e.g., interval or ratio) depending on how the measurement is used.

Increasing Levels of Measurement in the Meaningfulness of the Numbers

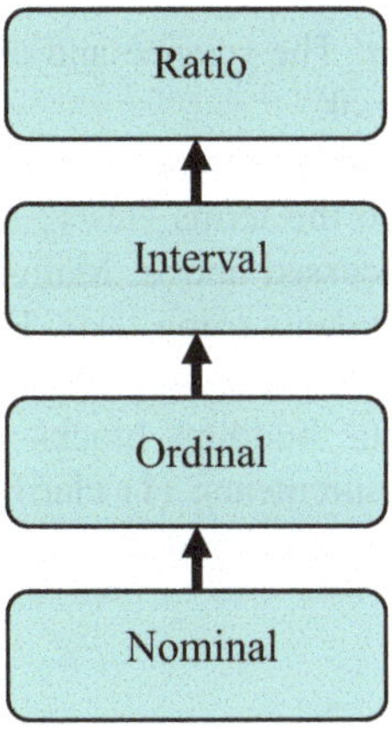

It can be seen that the four levels of measurement from nominal to ratio provides increasing power in the meaningfulness of the numbers used for measurement. If a measurement is at the ratio level, then comparisons between numbers both in terms of differences and in terms of ratios are meaningful. If a measurement is at the interval level, then comparisons between the numbers in terms of differences are meaningful. For ordinal measurements, only ordering can be inferred from the numbers, and not the actual distances between the numbers. Nominal level numbers do not provide much information in terms of "measurement" as defined in this book. For a comprehensive exposition on levels of measurement, see Khurshid and Sahai (1993).

Clearly, when one is developing a scale for measuring latent traits, it will be best if the numbers on the scale represent the highest level of measurement. However, in general, in measuring latent traits, there is no meaningful zero. It is difficult to construct an instrument to determine a total absence of a latent trait. So, typically for measuring latent traits, if one can achieve interval measurement for the scale constructed, the scale can provide more information than that provided by an ordinal scale where only rankings of objects can be made. Bearing these points in mind, Chap. 6 examines the properties of an ideal measurement in the psycho-social context.

The Process of Constructing Psycho-social Measurements

For physical measurements, typically there are well-known and well-tested instruments designed to carry out the measurements. Rulers, weighing scales and blood pressure machines are all examples of measuring instruments. In contrast, for measuring latent traits, there are no ready-made machines at hand, so we must first develop our "instrument". For measuring student achievement, for example, the instrument could be a written test. For measuring attitudes, the instrument could be a questionnaire. For measuring stress, the instrument could be an observation checklist. Before measurements can be carried out, we must first design a test or a questionnaire, or collect a set of observations related to the construct that we want to measure. Clearly, in the process of psycho-social measurements, it is essential to have a well-designed instrument. The science and art of designing a good instrument is a key concern of this book.

Before proceeding to explain about the process of measurement, we note that in the following, we frequently use the terms "tests" and "students" to refer to "instruments" and "objects" as discussed above. Many examples of measurement in this book relate to measuring students using tests. However, all discussions about students and tests are applicable to measuring any latent trait.

Wilson (2005) identifies four building blocks underpinning the process of constructing psycho-social measurements: (1) clarifying the construct, (2) developing test items, (3) gathering and scoring item responses, (4) producing measures, and then returning back to the validation of the construct in (1). These four building blocks form a cycle and may be iterative.

The key steps in constructing measures are briefly summarised below. More detailed discussions are presented throughout the book. In particular, Chap. 2 discusses defining the construct and writing test items. Chapter 3 discusses considerations in administering and scoring tests. Chapter 4 identifies key points in preparing item response data. Chapter 5 explains test reliability and classical test theory item statistics. The remainder of the book is devoted to the production of measures using item response modelling.

Define the Construct

Before an instrument can be designed, the construct (or latent trait) being measured must be clarified. For example, if we are interested in measuring students' English language proficiencies, we need to define what is meant by "English language proficiencies". Does this construct include reading, writing, listening and speaking proficiencies, or does it only include reading? If we are only interested in reading proficiencies, there are also different aspects of reading we need to consider. Is it just about comprehension of the language (e.g., the meaning of words), or about the "mechanics" of the language (e.g., spelling and grammar), or about higher-order cognitive processes such as making inferences and reflections from texts. Unless there is a clearly defined construct, we will not be able to articulate exactly what we are measuring. Different test developers will likely design somewhat different tests if the construct is not well-defined. Students' test scores will likely vary depending on the particular tests constructed. Also the interpretation of the test scores will be subject to debate.

The definition of a measurement construct is often spelt out in a document known as an assessment framework document. For example, the OECD PISA produced a reading framework document (OECD 2009) for the PISA reading test. Chapter 2 of this book discusses constructs and frameworks in more detail.

Distinguish Between a General Survey and a Measuring Instrument

Since a measuring instrument sometimes takes the form of a questionnaire, there has been some confusion regarding the difference between a questionnaire that seeks to gather separate pieces of information and a questionnaire that seeks to measure a central construct. A questionnaire entitled "management styles of hospital administrators" is a general survey to gather information about different management styles. It is not a measuring instrument since management styles are not being given scores from low to high. The questionnaire is for the purpose of finding out what management styles there are. In contrast, a questionnaire entitled "customer satisfaction survey" could be a measuring instrument if it is feasible to construct a satisfaction scale from low to high and rate the level of each customer's satisfaction. In general, if the title of a questionnaire can be rephrased to begin with "the extent to which....", then the questionnaire is likely to be measuring a construct to produce scores on a scale.

There is of course a place for general surveys to gather separate pieces of information. But the focus of this book is about methodologies for measuring latent traits. The first step to check whether the methodologies described in this book are appropriate for your data is to make sure that there is a central construct being measured by the instrument. Clarify the nature of the construct; write it down as "the extent to which ..."; and draft some descriptions of the characteristics at high and low levels of the construct. For example, a description for high levels of stress could include the severity of insomnia, weight loss, feeling of sadness, etc. A customer with low satisfaction rating may make written complaints and may not return. If it is not appropriate to think of high and low levels of scores on the questionnaire, the instrument is not likely a measuring instrument.

Write, Administer, and Score Test Items

Test writing is a profession. By that we mean that good test writers are professionally trained in designing test items. Test writers have the knowledge of the rules of constructing items, but at the same time they have the creativity in constructing items that capture students' attention. Test items need to be succinct but yet clear in meaning. All the options in multiple choice items need to be plausible, but they also need to separate students of different ability levels. Scoring rubrics of test items need to be designed to match item responses to different ability levels. It is challenging to write test items to tap into higher-order thinking. All of these demands of good item writing can only be met when test writers have been well trained. Above all, test writers need to have expertise in the subject area of what is being tested so they can gauge the difficulty and content coverage of test items.

Test administration is also an important step in the measurement process. This includes the arrangement of items in a test, the selection of students to participate in a test, the monitoring of test taking, and the preparation of data files from the test booklets. Poor test administration procedures can lead to problems in the data collected and threaten the validity of test results.

Produce Measures

As psycho-social measurement is about constructing measures (or, scores and scales) from a set of observations (indicators), the key methodology is about how to summarise (or aggregate) a set of data into a score to represent the measure on the latent trait. In the simplest case, the scores on items in a test, questionnaire or observation list can be added to form a total score, indicating the level of latent trait. This is the approach in classical test theory (CTT), or sometimes referred to as the true score theory where inferences on student ability measures are made using test scores. A more sophisticated method could involve a weighted sum score where different items have different weights when item scores are summed up to form the total test score. The weights may depend on the "importance" of the items. Alternatively, the item scores can be transformed using a mathematical function before they are added up. The transformed item scores may have better measurement properties than the raw scores. In general, IRT provides a methodology for summarising a set of observed ordinal scores into a measure that has interval properties. For example, the agreement ratings on an attitude questionnaire are ordinal in nature (with ratings 0, 1, 2, …), but the overall agreement measure we obtain through a method of aggregation of the individual item ratings is treated as a continuous variable with interval measurement property. Detailed discussions on this methodology are presented in Chaps. 6 and 7.

In general, IRT is designed for summarising data that are ordinal in nature (e.g. correct/incorrect or Likert scale responses) to provide measures that are continuous. Specifically, many IRT models posit a latent variable that is continuous and not directly observable. To measure the latent variable, there is a set of ordinal categorical observable indicator variables which are related to the latent variable. The properties of the observed ordinal variables are dependent on the underlying IRT mathematical model and the values of the latent variable. We note, however, that as the levels of an ordinal variable increases, the limiting case is one where the item responses are continuous scores. Samejima (1973) has proposed an IRT model for continuous item responses, although this model has not been commonly used.

We note, however, under other statistical methods such as factor analysis and regression analysis, measures are typically constructed using continuous variables. But item response functions in IRT typically link ordinal variables to latent variables.

Reliability and Validity

The process of constructing measures does not stop after the measures are produced. Wilson (2005) suggests that the measurement process needs to be evaluated through a compilation of evidence supporting the measurement results. This evaluation is typically carried out through an examination of reliability and validity, two topics frequently discussed in measurement literature.

Reliability

Reliability refers to the extent to which results are replicable. The concept of reliability has been widely used in many fields. For example, if an experiment is conducted, one would want to know if the same results can be reproduced if the experiment is repeated. Often, owing to limits in measurement precision and experimental conditions, there is likely some variation in the results when experiments are repeated. We would then ask the question of the degree of variability in results across replicated experiments. When it comes to the administration of a test, one asks the question "how much would a student's test score change should the student sit a number of similar tests?" This is one concept of reliability. Measures of reliability are often expressed as an index between 0 and 1, where an index of 1 shows that repeated testing will have identical results. In contrast, a reliability of 0 shows that a student's test scores from one test administration to another will not bear any relationship. Clearly, higher reliability is more desirable as it shows that student scores on a test can be "trusted".

The definitions and derivations of test reliability are the foundations of classical test theory (Gulliksen 1950; Novick 1966; Lord and Novick 1968). Formally, an observed test score, X, is conceived as the sum of a true score, T, and an error term, E. That is, $X = T + E$. The true score is defined as the average of test scores if a test is repeatedly administered to a student (and the student can be made to forget the content of the test in-between repeated administrations). Alternatively, we can think of the true score T as the average test score for a student on similar tests. So it is conceived that in each administration of a test, the observed score departs from the true score and the difference is called measurement error. This departure is not caused by blatant mistakes made by test writers, but it is caused by some chance elements in students' performance on a test. Defined this way, it can be seen that if a test consists of many items (i.e. a long test), then the observed score will likely be closer to the true score, given that the true score is defined as the average of the observed scores.

Formally, test reliability is defined as $\frac{Var(T)}{Var(X)} = \frac{Var(T)}{Var(T)+Var(E)}$ where the variance is taken across the scores of all students (see Chap. 5 on the definitions and derivations of reliability). That is, reliability is the ratio of the variance of the true scores over the variance of the observed scores across the population of students. Consequently, reliability depends on the relative magnitudes of the variance of the true scores and the variance of error scores. If the variance of the error scores is small compared to the variance of the true scores, reliability will be high. On the other hand, if measurement error is large, leading to a large variance of errors, then the test reliability will be low. From these definitions of measurement error and reliability, it can be seen that the magnitude of measurement error relates to the variation of an individual's test scores, irrespective of the population of respondents taking the test. But reliability depends both on the measurement error and the spread of the true scores across all students so that it is dependent on the population of examinees taking the test.

In practice, a reliability index known as Cronbach's alpha is commonly used (Cronbach 1951). Chapter 5 explains in more detail about reliability computations and properties of the reliability index.

Validity

Validity refers to the extent to which a test measures what it is claimed to measure. Suppose a mathematics test was delivered online. As many students were not familiar with the online interface of inputting mathematical expressions, many students obtained poor results. In this case, the mathematics test was not only testing students' mathematics ability, but it also tested familiarity with using online interface to express mathematical knowledge As a result, one would question the validity of the test, whether the test scores reflect students' mathematics ability only, or something else in addition to mathematics ability.

To establish the credibility of a measuring instrument, it is essential to demonstrate the validity of the instrument. Standards for Educational and Psychological Testing (AERA, APA, NCME 1999) (referred to as the Standards document hereafter) describe several types of validity evidence in the process of measurement. These include:

Evidence based on test content

Traditionally, this is known as content validity. For example, a mathematics test for grade 5 students needs to be endorsed by experts in mathematics education as reflecting the grade 5 mathematics content. In the process of measurement, test content validity evidence can be collected through matching test items to the test specifications and test frameworks. In turn, test frameworks need to be matched to the purposes of the test. Therefore documentations from the conception of a test to the development of test items can all be gathered as providing the evidence of test content validity.

Evidence based on response process

In collecting response data, one needs to ensure that a test is administered in a "fair" way to all students. For example, there are no disturbances during testing sessions and adequate time is allowed. For students with language difficulties or other impairments, there are provisions to accommodate these. That is, there are no extraneous factors influencing student results in the test administration process. To collect evidence for the response process, documentations relating to test administration procedures can be presented. If there are judges making observations on student performance, the process of scoring and rater agreement need to be evaluated.

Evidence based on internal structure

The relationship (inter-correlation) among test items gives us some indication of the degree to which the test items "hang together" to reflect a single construct. For example, if we construct a questionnaire to measure extraversion/introversion in personality, we may find that "shyness" does not relate highly to "preference to be alone", but we may have hypothesised a close relationship when designing the instrument. The data of item responses from instrument administrations allow us to check whether the items tap into one construct or multiple constructs. We can then match the theoretical construct defined in the beginning of the measurement process and the empirically established constructs. This match will provide evidence of construct validity.

Standards for Educational and Psychological Testing (AERA, APA, NCME 1999) also include validity evidence based on relations to other variables, and evidence based on consequences of testing. We refer the readers to the Standards document. In summary, validity evidence needs to be collected "along the way" of constructing measures, starting from defining the construct to producing measurement scores. The Standards document places Validity as the opening chapter in the document, emphasising its importance in psycho-social measurement. A detailed discussion of validity is beyond the scope of this book. Interested readers are referred to Messick (1989) and Lissitz (2009) for further information.

Graphical Representations of Reliability and Validity

Frequently, a graphical representation is used to explain the differences between reliability and validity. Figure 1 shows such a graph.

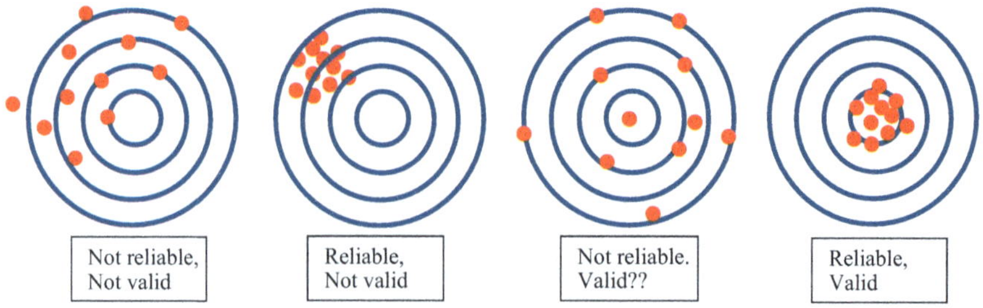

Fig. 1.1 Graphical representations of the relationship between reliability and validity

Figure 1 represents reliability and validity in the context of target shooting where reliability is represented by the closeness of the scores under repeated attempts by a shooter, and validity is represented by how close the average location of the scores is to the centre of the target. However, in the contexts of psychological testing, if an instrument does not have satisfactory reliability, one typically cannot claim validity. That is, validity requires that instruments are sufficiently reliable. So the third picture in Fig. 1 does not have validity because the reliability is low.

Summary

This chapter introduces the idea of educational and psychological measurements by contrasting it with physical measurements. The main definitions of measurement relate to the assignment of numbers to objects to order, or quantify, some attributes of objects. Constructed measures have different levels in terms of the amount of information conveyed by the measured scores: nominal, ordinal, interval and ratio. Typically in educational and psychological measurements, we aim for ordinal or interval measures.

The process of constructing measures consists of a number of key steps: defining the construct, developing instruments, administering instruments and collecting data, producing measures. After measures are produced, there is an evaluation process of the measurement through an examination of reliability and validity of the instrument.

Discussion Points

1. Discuss whether latent variables should have a meaningful zero and why it may be difficult to define a zero.
2. Given that there could be a meaningful zero for test scores where zero means a student answered all questions incorrectly, are test scores ordinal, interval or ratio variables? If test scores are used as measures of an underlying ability, what level of measurement are test scores?
3. Is the following a "measurement" instrument? If so, what is the construct being measured?

Car Survey

"What characteristics led to your decision for the specific model?"

Tick four categories

	Customer 1	Customer 2	Customer 3	Customer 4
Economy	✓			✓
Handling		✓	✓	✓
Interior design	✓	✓		
Exterior design	✓		✓	✓
Reliability		✓		
Price			✓	✓
Comfort	✓			
Safety		✓	✓	

4. Is the following a "measurement" instrument? If so, what is the construct being measured?

Taxi Survey

Rating taxi rides

Melb airport to kew	Taxi 1	Taxi 2	Taxi 3	Taxi 4
Comfortable temperature	✓	✓	✗	✓
Driver's certificate displayed	✓	✓	✗	✗
Uniform correct	✗	✓	✗	✓
Driver presentation	✓	✓	✗	✓
Pleasant odour	✓	✗	✗	✓
Internal cleanliness	✓	✓	✗	✓
External cleanliness	✓	✗	✗	✓
Vehicle handling	✓	✓	✗	✓
Driver quality	✓	✗	✗	✓
Correct change	✗	✗	✗	✓
Politeness	✓	✓	✗	✓
Peak time	✗	✗	✓	✓
Metered fare	$47.10	$48.40	$50.40	$51.00

5. Messick (1989) provided a definition of validity as "an integrated evaluative judgment of the degree to which empirical evidence and theoretical rationales support the adequacy and appropriateness of inferences and actions based on

test scores or other modes of assessment." Compare and contrast this definition of validity with what we have discussed in this chapter. Do you think this is a good definition of validity? Provide reasons for your answers.

Exercises

Q1. The following are some data collected in SACMEQ (Southern and Eastern Africa Consortium for Monitoring Educational Quality, UNESCO-IIEP 2004). For each variable, state whether the numerical coding as shown in the boxes provides nominal, ordinal, interval or ratio measures?

1. *PENGLISH*
   Do you speak English outside school?
   (Please tick only one box.)
   - ☐ (1) Never
   - ☐ (2) Sometimes
   - ☐ (3) All of the time

2. *XEXPER*
   How many years ***altogether*** *have you been teaching?*
   (Please round to '1' if it is less than 1 year.)
   ☐☐ years

3. *PCLASS*
   Which Standard 6 class are you in this term?
   (Please tick only one box.)

6A	6B	6C	6D	6E	6F	6G	6H	6I	6J	6K	6L
☐ (01)	☐ (02)	☐ (03)	☐ (04)	☐ (05)	☐ (06)	☐ (07)	☐ (08)	☐ (09)	☐ (10)	☐ (11)	☐ (12)

4. *PSTAY*
   Where do you stay during the school week?
   (Please tick only one box.)
   - ☐ (1) In my parents'/legal guardian's home
   - ☐ (2) With other relatives or another family
   - ☐ (3) In a hostel/boarding school accommodation
   - ☐ (4) Somewhere by myself or with other children

Q2. Which questionnaire titles in the following list would appear to be about "measurement" (as opposed to a survey)?

	Sports familiarity questionnaire
	What are the different management structures of government departments in Victoria?
	Where can senior citizens find help?
	How happy are you?
	Proficiency in statistics
	Finding out your stress level

Q3. On a mathematics test of 40 questions, Jenny got a score of 14. Eric got a score of 28. Mary got a score of 30.

We can be **reasonably confident** to conclude that (write Yes or No in the space provide)

1. Jenny is not as good in mathematics as Eric and Mary are. [ ]
2. Mary is better at mathematics than Eric is. [ ]
3. Eric got twice as many questions right as Jenny did. [ ]
4. Eric's mathematics ability is twice Jenny's ability. [ ]

Q4. A movie guide rates movies by showing a number of stars. For example, a movie with 3-and-a-half stars is not as good as a movie with 4 stars (★★★★☆).

What is the most likely measurement level provided by this kind of ratings?

	Nominal
	Ordinal
	Interval
	Ratio

Q5. In the context of educational testing, the term "measurement error" usually refers to

	The variation in a student's scores, should similar tests be administered
	The number of errors a student makes in a test
	Errors in the questions of a test, e.g., incorrect questions
	Errors in the processing of data, e.g., marker error, data entry error
	Careless mistakes made by anyone (e.g., students, test setters and/or markers)

Q6. In the context of educational testing, test reliability refers to

	The degree to which the test questions reflect the construct being tested
	The degree to which a test is error-free or error-prone
	The degree to which test scores can be reproduced if similar tests are administered
	The extent to which a test is administered to candidates (e.g., the number of test takers)

Q7. A student with limited proficiencies in English sat a Year 5 mathematics test and obtained a poor score due to language difficulties. Is this an issue related to test reliability or validity?

Q8. In a Grade 5 spelling test, there are 20 words. This is a very small sample of all the words Grade 5 students should know. If the test is used to measure students' spelling proficiency in general, which of the following best describes the likely problems with this test?

	There will be a problem with reliability, but NOT validity
	There will be a problem with validity, but NOT reliability
	There will be a problem with BOTH reliability and validity
	We cannot judge whether there will be a problem with reliability or validity

References

AERA, APA, NCME (1999) Standards for educational and psychological testing. American Educational Research Association, Washington

Allen MJ, Yen WM (1979) Introduction to measurement theory. Brooks/Cole Publishing Company, Monterey, California

Cronbach LJ (1951) Coefficient alpha and the internal structure of tests. Psychometrika 16(3): 297–334

Gulliksen H (1950) Theory of mental tests. Wiley, New York

Khurshid A, Sahai H (1993) Scales of measurements: an introduction and a selected bibliography. Qual Quant 27:303–324

Lissitz RW (ed) (2009) The concept of validity. Revisions, new directions, and applications. Information Age Publishing, Inc., Charlotte

Lord FM, Novick MR (1968) Statistical theories of mental test scores. Addison-Wesley, Reading

Messick S (1989) Validity. In: Linn R (ed) Educational measurement, 3rd edn. American Council on Education/Macmillan, Washington, pp 13–103

Novick MR (1966) The axioms and principal results of classical test theory. J Math Psychol 3 (1):1–18
Nunnally JC, Bernstein IH (1994) Psychometric theory. McGraw-Hill Book Company, New York
OECD (2009) PISA 2009 Assessment framework—key competencies in reading, mathematics and science. Retrieved 28 Nov 2012, from http://www.oecd.org/pisa/pisaproducts/44455820.pdf
Samejima F (1973) Homogeneous case of the continuous response model. Psychometrika 38: 203–219
Stevens SS (1946) On the theory of scales of measurement. Science 103:667–680
Thomson M (2003) The application of Rasch scaling to wine judging. Int Edu J 4(3):201–223
UNESCO-IIEP (2004) Southern and Eastern Africa Consortium for monitoring educational quality (SACMEQ) Data Archive. See http://www.sacmeq.org/data_archive.htm
Wilson M (2005) Constructing measures: an item response modeling approach. Lawrence Erlbaum Associates, Mahwah
Wright BD, Masters GN (1982) Rating scale analysis: Rasch measurement. Mesa Press, Chicago

Further Reading

Bartholomew DJ (ed) (2006) Measurement. Volume 1. Sage Publications Ltd.
Brennan RL (ed) (2006) Educational measurement, 4th edn. Praeger publishers, Westport
Furr RM, Bacharach VR (2008) Psychometrics: an introduction. Sage Publications Ltd, Thousand Oaks
Thorndike RM, Thorndike-Christ T (2010) Measurement and evaluation in psychology and education, 8th edn. Pearson Education, Upper Saddle River
Walford G, Tucker E, Viswanathan M (eds) (2010) The SAGE handbook of measurement. SAGE publications Ltd., Thousand Oaks

Chapter 2
Construct, Framework and Test Development—From IRT Perspectives

Introduction

In Chap. 1, the terms "latent trait" and "construct" are used to refer to the psycho-social attributes that are of interest to be measured. How are "constructs" conceived and defined? Can a construct be any arbitrarily defined concept, or does a construct need to have specific properties in terms of measurement? The following is an example to stimulate some thoughts about constructs.

There is an Australian radio station RPH (Radio for the Print Handicapped) that read newspapers and books aloud to listeners. To demonstrate the importance of this radio station, listeners of RPH are constantly reminded that "1 in 10 in our population cannot read print". This statement raises an interesting question. That is, if an instrument is developed to measure people's ability to read print, how would one go about doing it? And how does this differ from the 'reading abilities' we are accustomed to measure through achievement tests?

To address these questions, the starting point is to clearly define the "construct" of such a measuring instrument. Loosely speaking, the construct can be defined as "what we are trying to measure". We need to be clear about what it is that we are trying to measure before test development can proceed.

In the case of the RPH radio station, one's first impression is that this radio station is for vision-impaired people. Therefore, to measure the ability to read print for the purpose of assessing the targeted listeners of RPH is to measure the degree of vision impairment of people. This, no doubt, is an overly simplified view of the services of RPH. In fact, RPH can also serve those who have low levels of reading ability and do not necessarily have vision impairment. Furthermore, people with low levels of reading achievement but also a low level of the English language would not benefit from RPH. For example, immigrants may have difficulties to read newspapers, but they will also have difficulties in listening to broadcasts in English. There are also people who spend a great deal of time in cars and traffic jams, and who find it easier to "listen" to newspapers than to "read" newspapers even though

M. Wu et al., *Educational Measurement for Applied Researchers*,
DOI 10.1007/978-981-10-3302-5_2

these people have high levels of reading ability. Thus the definition of "the ability to read print", for RPH, is not straightforward to define. What we may want to measure is the degree to which a person finds it useful to have print materials read to them. If ever an instrument is developed to measure this, the construct needs to be carefully examined.

Linking Validity to Construct

The above example illustrates that, in clarifying a construct, the purposes of the measurement need to be considered. Generally, the notion of a construct in psycho-social measurements may be somewhat fluid in that definitions are shaped depending on the contexts and purposes of the measurements. For example, there are many different definitions for a construct called "reading ability", depending on the contexts in which measures are made. In contrast, measurements in the physical world often are attached to definitions based on scientific theories and the measures are more clearly defined.

In shaping a psycho-social construct, we need to first consider validity issues. That is, the inferences made from measurement scores and the use of these scores should reflect the definition of the construct. Consequently, when constructs are defined, one should clearly anticipate the ways the scores are intended to be used, or at least clarify to the users of the instrument the inferences that can be drawn from the scores.

There are many different purposes for measurement. A classroom teacher may set a test to measure the extent to which students have learned two science topics taught in a semester. In this case, the test items will be drawn from the material that was taught, and the test scores will be used to report the proportion of knowledge/skills students have acquired from class instructions in that semester. The construct of this test will be related to how well students grasp the material that was taught in class. The test scores will not be used to reflect general science abilities of the students.

In developing state-wide achievement tests, it is often the case that the content, or curriculum coverage, is used to define the construct for the test. Therefore one might develop a mathematics test based on the Curriculum Standards Framework or other official documents. That is, what is tested is the extent to which students have attained the intended mathematics curriculum. Any other inferences made about the test scores such as the suitability for course entry, employment, or general levels of mathematics literacy, will need to be treated with caution.

What if one wants to make inferences about students' abilities beyond the set of items in a test? What assumptions will need to be made about the test and test items so one can provide some generalisations of students' scores? Consider the PISA (Programme for International Student Assessment) tests, where the constructs are not based on school curricula. Can one make statements that the PISA scores reflect the levels of general mathematics, reading and science literacy? What are the conditions under which one can make inferences beyond the set of items in a test? Clearly, the evaluation of reliability and validity discussed in Chap. 1 plays an important role. In this chapter, we will take a look at the role Item Response Theory (IRT) plays in relation to defining a construct.

Construct in the Context of Classical Test Theory (CTT) and Item Response Theory (IRT)

Under classical test theory and item response theory, there are theoretical differences in the meaning of the construct, although for all practical purposes the distinction is not important. Under the approach of the classical test theory, inferences made are about a person's score on a test. While there is no explicit generalisation about the level of a "trait" that a person might possess, the 'true score' defined CTT reflects the construct we are measuring. Under the notion of 'parallel tests' in CTT, a construct can be construed implicitly through the test items in these parallel tests. In contrast, under the approaches of IRT, there is an explicit latent trait defined in the model. An instrument sets out to measure the level of the latent trait in each individual. The item responses and the scores of a student reflect the level of this trait of the student. The trait is said to be "latent", because it is not directly observable. Figure 2.1 shows a latent trait model under the IRT approach.

In Fig. 2.1, the latent variable is the construct to be measured. Some examples of a latent variable could be proficiency in geometry, asthma severity, support for an initiative, familiarity with sport, etc. Since one cannot directly measure a latent variable, "items" will need to be devised to tap into the latent variable. A person's

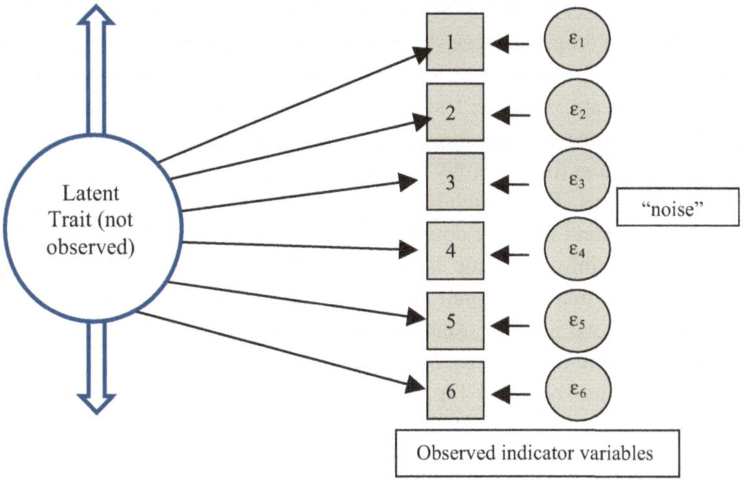

Fig. 2.1 Latent variables and indicator (observable) variables

response on an item is observable. In this sense, the items are sometimes known as "observed indicator variables" or "manifest variables". Through a person's item response patterns, some inferences can be made about a person's level on the latent variables. The items represent small concepts based on the bigger concept of the latent variable. For example, if the latent variable is proficiency in geometry, then the items are individual questions about specific knowledge or skills in geometry.

The arrows (from the latent variable to the observed indicators) in Fig. 2.1 indicate that the level of the latent variable influences the likely responses to the items. It is important to note the direction of the arrows. That is, the item response pattern is driven by the level of the latent variable. It is not the other way round that the latent variable is defined by the item responses. For example, the consumer price index (CPI) is defined as the average price of a fixed number of goods. If the prices of these goods are regarded as items, then the average of the prices of these items defines CPI. In this case, CPI should not be regarded as a latent variable. Rather, it is an index defined by a fixed set of some observable entities. We cannot change the set of goods and still retain the same meaning of CPI. In the case of IRT, since the level of the latent variable determines the likelihood of the item responses, the items can be changed, for as long as all items tap into the same latent variable, and we will still be able to measure the level of the latent variable.

In Fig. 2.1, the symbol "ε" indicates "noise" in the sense that items can possibly be influenced by factors other than the latent variable. It is clearly undesirable to have large "noises", since these interfere with the measurement of the latent trait. The CTT notion of reliability discussed in Chap. 1 relates to the amount of "noise" the item scores have. The more noise there is, the lower the reliability. Through item analysis, the relative amount of noise for each item can be identified to determine the degree to which an item taps into the latent trait being measured.

Under classical test theory, only the right-hand side of the picture (observed indicators) of Fig. 2.1 is involved, as shown in Fig. 2.2.

Consequently, under classical test theory, inferences about the score on this set of items (and scores on parallel tests) can be made. The construct being measured is implicitly represented by the 'true score' (defined as the average of test scores of parallel tests). We can exchange test items in a test in the context of parallet tests.

Under item response theory, the notion of CTT parallel tests is replaced by an explicitly defined latent trait whereby any item tapping into the latent trait can be used as potential test items. Consequently, we can exchange items in the test and still measure the same latent trait. Of course, this relies on the assumption that the items used indeed all tap into the same latent trait. This assumption needs to be tested before we can claim that the overall performance on the test reflects the level of the latent trait. That is, we need to establish whether arrows in Fig. 2.1 can be placed from the latent variable to the items. It may be the case that some items do not tap into the latent variable, as shown in Fig. 2.3. As IRT has an underlying mathematical model to predict the likelihood of the item responses, statistical tests of fit can be constructed to assess the degree to which responses of an item "fit" the IRT model. Such fit tests provide information on the degree to which individual items are indeed tapping into the latent trait. Chapter 8 discusses about IRT fit statistics.

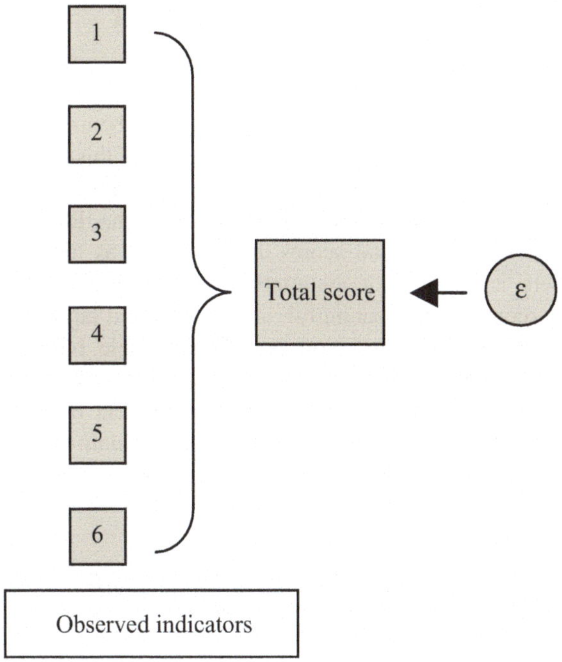

Fig. 2.2 Model of classical test theory

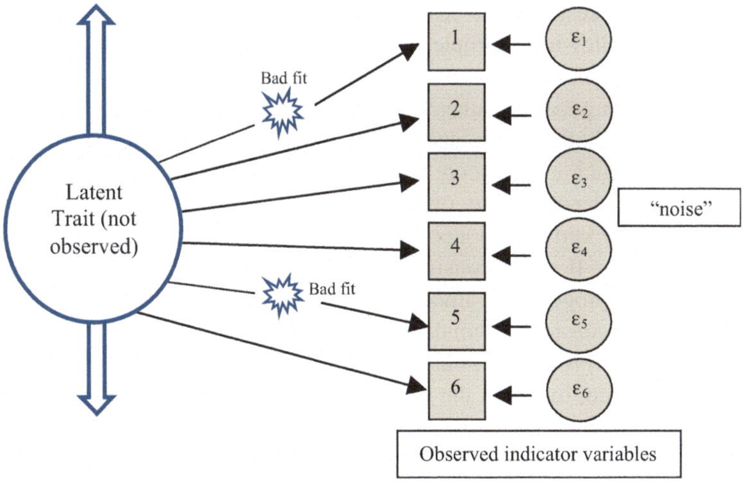

Fig. 2.3 Test whether items tap into the latent variable

Unidimensionality in Relation to a Construct

The IRT model shown in Fig. 2.1 shows that there is one latent variable and all items tap into this latent variable. This model is said to be unidimensional, in that there is ONE latent variable of interest, and the level of this latent variable is the focus of the measurement. If there are multiple latent variables to be measured in one test, and the items tap into different latent variables, the IRT model is said to be multidimensional. Whenever total scores are computed as the sum of individual item scores, there is an implicit assumption of unidimensionality. That is, for aggregated item scores to be meaningful, all items should tap into the same latent variable. Otherwise, an aggregated score is un-interpretable, because the same total score for students A and B could mean that student A scored high on latent variable X, and low on latent variable Y, and vice versa for student B, when there are two different latent variables involved in the total score. Multidimensional IRT models are discussed in Chap. 15.

The Nature of a Construct—Psychological Trait or Arbitrarily Defined Construct?

The theoretical notion of latent traits as shown in Fig. 2.1 seems to suggest that there exists distinct latent traits (e.g., "abilities") within each person, and the construct must reflect one of these distinct abilities for the item response model to hold. This is not necessarily the case in practice.

Consider the following example. Reading and mathematics are considered as different latent variables in most cases. That is, a student who is good at reading is not necessarily also good at mathematics. So in general, one would not administer one test containing both reading and mathematics items and compute a total score for each student. Such a total score would be difficult to interpret.

However, consider the case of mathematical problem solving, where each problem requires a certain amount of reading and mathematics proficiencies to arrive at an answer. If a test consists of problem solving items where each item requires the same "proportional amount" of reading ability and mathematics ability, the test can still be considered "unidimensional", with a single latent variable called "problem solving". From this point of view, whether a test is "unidimensional" depends on the extent to which the items are testing the same construct, where the construct can be defined as a composite of abilities (Reckase et al. 1988).

In short, latent variables do not have to correspond to distinct "traits" or "abilities" as we commonly perceive. Latent variables are constructs defined by the researcher to serve his/her purpose of measurement.

Practical Considerations of Unidimensionality

In practice, one is not likely to find two items that test exactly the same construct, since all items require different composite abilities. So all tests with more than one item are "multidimensional" to different degrees. For example, the computation of "7 × 9" may involve quite different cognitive processes to the computation of "27 + 39". To compute "7 × 9", it is possible that only recall is required for those students who were drilled on the "Multiplication Table". To compute "27 + 39", some procedural knowledge is required. However, one would say that these two computational items are still closer to each other for testing the construct of basic computational ability than, say, solving a crossword puzzle. So in practice, the dimensionality of a test should be viewed in terms of the practical utility of the use of the test scores. For example, if the purpose of a test is to select students for entering into a music academy, then a test of "music ability" may be constructed. If one is selecting an accompanist for a choir, then the specific ability of piano playing may be the primary focus. Similarly, if an administrative position is advertised, one may administer a test of "general abilities" including both numeracy AND literacy items. If a company public relations officer is required, one may focus only on literacy skills. That is, the degree of specificity of a test depends on the practical utility of the test scores.

Theoretical and Practical Considerations in Reporting Sub-scale Scores

In achievement tests, there is often a problem of deciding how test scores should be reported in terms of cognitive domains. Typically, it is perceived to be more informative if a breakdown of test scores is given so that one can report on students' achievement levels in sub-areas of cognitive domains. For example, a mathematics test is often reported by a total score to reflect an overall performance on the whole test, and also by performances on mathematics sub-strands such as Number, Measurement, Space, Data, etc. Few people query about the appropriateness of such reporting, as this is how mathematics is specified in school curriculum. However, when one considers reporting from an IRT point of view, there is an implicit assumption that whenever sub-scales are reported, the sub-scales relate to different latent traits. Curriculum specifications, in general, take no explicit consideration of latent traits. Furthermore, since sub-scale level reporting implies that the sub-scales cannot be regarded as measuring the same latent trait, it will be theoretically incorrect to combine the sub-scales as one measure of a single latent trait. This theoretical contradiction, however, is generally ignored in practice. One may argue that, since most cognitive dimensions are highly correlated (e.g., Adams and Wu

2002), one may still be able to justify the combination of sub-scales within a subject domain to obtain an aggregate score representing students' proficiency in the subject domain.

Summary About Constructs

In summary, the clarification of the construct is essential before test construction. It is a step towards establishing what is being measured. Furthermore, if we want to make inferences beyond students' performances on the set of items in a test, additional assumptions about the construct need to be made. In the case of IRT, we begin by relating the construct of a test to some latent trait, and develop a framework to provide a clear explication of this latent trait.

It should be noted that there are two sides of the coin that need to be kept in mind. First, no two items are likely measuring exactly the same construct. If the sample size of test takers is large enough, all items will show misfit when tested for unidimensionality (see Chap. 8 for details). Second, while it is impossible to find items that measure the same construct, cognitive abilities are highly correlated so that in practice what one should be concerned with is not whether a test is unidimensional but whether a test is sufficiently unidimensional for the purposes of the use of the tests. Therefore, it is essential to link the construct to validity issues in justifying the fairness of the items in relation to how the test scores are used.

Nevertheless, while the assumption of unidimensionality is always only approximately satisfied, one should always aim to achieve it. Otherwise, there will be an instrument with items tapping into different constructs, and we will no longer be able to attach meanings to test scores. In that case, the instrument is no longer about measurement. Under this circumstance, general survey analysis should be used instead of methodologies for measurement.

Before closing the discussions on constructs and unidimensionality, one note should be made about the comparisons between classical test theory and item response theory. While IRT provides a better model for measurement as it begins with hypothesising a latent trait in contrast to CTT which focuses on the test items specific to a test, CTT still holds a notion that the test score reflects a measure on a construct defined by "similar tests" to the current test. CTT statistics such as test reliability and item discrimination indices also help with building an instrument with items correlated with each other (the notion of internal consistency). So, while there are theoretical differences between IRT and CTT as described in previous sections, in practice, both IRT and CTT help us with building a good measuring instrument. Consequently, CTT and IRT should be used hand-in-hand in a complementary way, and one should not discard one approach for another. See Chap. 5 for further information on CTT.

Table 2.1 Percentages of items for a hypothetical mathematics test

Skill content	Knowledge	Application	Reasoning	Row total
Number	5	10	10	25
Space	2	10	8	20
Data	5	15	5	25
Measurement	8	15	7	30
Column total	20	50	30	100

Frameworks and Test Blueprints

Given the importance of constructs in measurement, the first step in measurement is to provide a detailed description of the construct to be measured. Such descriptions are usually written in documents typically known as assessment frameworks and test blueprints. An assessment framework should cover the purpose of the assessment, the target population to be assessed, assessment methods, and mostly importantly, the definition of the construct to be measured and the content to be covered in the assessment. Typically, an assessment framework is written by subject matter experts, with contributions from all stakeholders of the assessment. In describing the construct, different aspects/dimensions associated with the construct may be identified. For example, for a mathematics achievement test, the curriculum content strands (e.g., algebra, number, geometry) can form one dimension while cognitive skills (e.g., recall, procedural knowledge, reasoning) can form another dimension. These two dimensions can be further divided into sub-strands to ensure coverage of the content domain.

Test blueprints are documents with specifications of item characteristics such as item format (multiple-choice or constructed response), range of item difficulties, percentages of items under each construct dimension and sub-dimensions. Table 2.1 shows an example of the percentages of items for a mathematics assessment.

Test specifications will be decided by balancing considerations with regard to the purpose of the test, characteristics of students, content coverage, assessment duration and other constraining factors.

Taken as a whole, the assessment framework plays the role as the overall plan that specifies how the test will be developed. Assessment frameworks also serve as monitoring tools for the content validity of the tests. Examples of assessment frameworks and test blueprints can be found from large-scale surveys such as PISA and TIMSS (e.g., OECD 2013; Mullis et al. 2009).

Writing Items

There are many reference materials on the science and art of writing test items (e.g., Osterlind 2002; Downing and Haladyna 2006). In this book, we will only describe briefly the key topics of item writing.

When developing a measuring instrument, it is most important to follow the framework and test blueprint documents. These documents ensure that the instrument contains balanced content and difficulty levels, as well as reflecting the constructs being measured. While one needs to bear in mind that items should all tap into the same construct, items ought to be independent in the sense that there are not the same question asked in slightly different ways, since we want each item to provide an independent and additional piece of information about a test taker. Further, the answer to a question should not be dependent on the answers for previous questions.

Item Format

There has been much debate about the pros-and-cons of multiple-choice items over constructed-response items. Clearly, when an item has the multiple-choice format, there is a possibility of guessing the correct answer. For this reason, constructed-response items typically have higher discrimination indices than multiple-choice items. There is also an issue with face validity for multiple-choice items, as typically in real-life, we need to solve problems and there is rarely a list available to us with the correct answer among the list. Nevertheless, multiple-choice items can be machine-scored easily, while constructed-response items often need human scorers, leading to increased cost. For open-ended items where markers are required to score the responses, marker agreements are not always good. Variations among marker harshness also contribute to the unreliability of test scores. Weighing up these considerations, multiple-choice items are still cost-effective to provide data for measurement.

There is one note of caution about using the multiple-choice format. Since the correct answer is among the list presented to the test taker, the cognitive processes for solving a problem may be changed from those of a constructed-response item. For example, the following is an item intended to test students' ability to solve an equation:

Given $-3x + 16 = -14$,
x equals
A. -1.5
B. 1.5
C. 10
D. -10

Since the correct answer is among the list, one can simply substitute the four possible answers in the equation to check which one satisfies the equation. That is, instead of re-arranging the terms in the equation to solve for x, the substitution strategy can be used. This item is likely to be easier than the following constructed-response item where the correct answer is not shown to the test taker:

Given $-3x + 16 = -14$
x equals ______

The following is another example.

Which of the following is the capital city of Ireland?
A. Dublin
B. Edinburgh
C. London
D. Oslo

The process of elimination can be used to arrive at the correct answer if the test taker can eliminate some wrong answers and make it easier to choose the correct answer. The constructed-response item shown below is likely to be more difficult.

The capital city of Ireland is ______

Consequently, a multiple-choice item may be less difficult than an open-ended item, even if the content of the item is the same, not only because of the possibility of guessing, but because a number of strategies can be used to obtain the correct answer. Multiple-choice items also tend to have lower discrimination indices because factors other than the intended latent trait also contribute to the chance of obtaining the correct answer.

As technology is advancing at a great pace and tests can be delivered by the computers, it becomes more feasible to administer closed constructed-response items such as the two examples shown above. The computer can score written responses for these two short-answered questions. Thus, wherever possible, machine score-able constructed-response items are preferable to multiple-choice items for increasing item discrimination and overall test reliability.

Number of Options for Multiple-Choice Items

Questions are constantly raised about the number of options required for a multiple-choice item. Should there be the same number of options for all multiple-choice items? Should there be at least four options for each item? There are no definite answers to these questions. Clearly, when there are fewer options, the chance of guessing the correct answer is greater. On the other hand, if some options are "unattractive" and few test takers choose them, it is pointless to include these options.

Depending on the context of a question, sometimes there is a set of "natural" options for an item. For example, the following item has seven "natural" options:

> If August 6 is a Monday, what day of the week is August 18?

In contrast, the following item has three "natural" options:

> If two classes of students sit a test, where Class 1 is a class of high achievers and Class 2 is a class of students with varying abilities, will the test reliability for Class 1 be higher, lower or the same as for Class 2?

We may be able to find a fourth option for the above item, but few will likely choose it. In general, it is a waste of time to come up with a fixed number of options if some options are clearly irrelevant to the question. It will be better to let the context of a question determine the best number of options rather than to make a rule for a fixed number of options for all items. While we need to be mindful of guessing, it may be a waste of test writers' and test takers' time to include many implausible options.

How Many Items Should There Be in a Test?

Clearly, the more information is gathered about a test taker, the more reliable the measure will be. A five-item mathematics test will not provide a very reliable measure of a person's proficiency in mathematics as compared to a 50-item test. Strictly speaking, it is not the number of items in a test that matters; it is the number of total score points of a test. For example, if a test has four questions each of which has a score range between 0 and 5 so the maximum score for the test is 20 score points, this four-item test provides similar amount of information as a test with 20 dichotomously scored items.

The measurement error associated with a test score decreases as the number of score points of the test increases. What magnitude of measurement error is acceptable? Unfortunately, there is not a simple answer. It depends on the purposes for which the test scores are used. If the purpose of the test is to determine whether a student has an average, above average or below average performance in mathematics, a 40-item test (or a 40 score-point test) will be sufficient to provide that information. If the purpose of the measurement is to estimate the growth of a student in one year, then two 40-item tests one year apart will not provide very accurate growth measure. This is because the expected magnitude of yearly growth is small as compared to the measurement errors of 40-item tests. Chapter3 further discusses measurement errors.

While we are not able to provide a definitive recommendation on the number of items (and score points) in a test, the following are some guidelines. For a typical test of about 40 questions taken in one-hour of test time, the measurement errors are rather large and the scores can only be used for low-stakes purposes. That is, such tests provide indicative information about whether a student is struggling or doing well. If a test is used for entrance examinations or for awarding high-stakes certification (e.g., qualification for a practising physician), then much longer tests are required (Wu 2010).

On the other hand, if the purpose of a test is for providing system level information such as the proficiency levels of students in a state or in a country, then there need to be many test items to cover the whole curriculum, but each student can take only a small portion of the test (say, 10 items). Reliable information can still be obtained at the system level. Chapter 3 discusses about test design and elaborates on different ways tests can be arranged and delivered to students for different purposes.

Scoring Items

It may seem appropriate that the correct answer to an item should be given a score of 1 and an incorrect answer a score of 0. Consider the following item.

What is the area of a rectangular room measuring 4 m by 6 m?

How should answers such as "24 m^2", "24", "24 m", "10 m" be scored? Clearly, "24 m^2" should get a score of 1 and "20 m" should get a score of 0. How about "24", 24 m"? We may argue that "24" is a correct answer so should be given a score of 1; but "24 m" is technically incorrect with the unit of area, so it should be given a score of 0. This scoring scheme, however, is not consistent with measurement considerations. From a psychometric viewpoint, the score on an item should reflect the test taker's level on the latent trait. With respect to this item, a person answering "24 m" is likely to be higher on the scale than a person answering "10 m". But yet we give them the same score (0 score). A person answering "24" is not necessarily better than a person answering "24 m", but yet we give the former a

"1" and the latter a "0" score. From a psychometric viewpoint, it will be better to give all three answers ("24 m^2", "24", "24 m") a score of 1, even though there are some technically incorrect answers, to distinguish these students from those who really have no idea about the formula for computing the area of a rectangle. Measurement is about making the best prediction of a person's level on a scale. The scoring scheme of responses to questions should reflect this level on the latent trait.

If subject experts have objections about giving a score of "1" when there are technically incorrect answers, then the test item should be revised to prevent such technically incorrect answers. For example, we may provide the unit of area and ask students to find the number only. If we are interested in testing the use of units, we may design a question just about units of area.

Awarding Partial Credit Scores

In the example above, when some item responses are "better" than other responses, there is a possibility of providing partial credit scoring. For example, the answer "24 m^2" may be given a score of 2, "24" and "24 m" a score of 1, and "10 m" a score of 0. We may also consider giving scores of 1, 0.5 and 0 to these three sets of responses. How does one decide on the scoring and what are the considerations in providing partial credit scores?

First, when partial credit scorings are used, increasing scores must correspond to increasing underlying level of latent trait. For example, the following item is from an instrument that measures extraversion/introversion of a person.

> You have been invited by a good friend to a party but you do not know anyone other than the host at this party. What would you do? (check one box only).
>
> - Eagerly accept the invitation and mix with most people at the party.
> - Turn down the invitation even though the host is your good friend.
> - Accept the invitation but stick next to you friend throughout the party.
> - Accept the invitation and hang around with a small group of people with similar interest.

The four options correspond to different levels on the extraversion/introversion scale. Respondents choosing the first option are likely to be high (more extraverted) on the scale while respondents choosing the second option are likely to be low on the scale. Those respondents choosing the third and fourth options are likely to be in the middle of the scale. One might propose a scoring scheme of 3, 0, 1, 2 for the four options respectively, or, 2, 0, 1, 1 if the third and fourth options are deemed similar. At the stage of the proposal of a scoring scheme, a hypothesis is made about the partial credit order based on expert judgement of how the responses match levels of the latent trait. After data have been collected, the data analysis will help us verify the scoring schemes (See Chaps. 9 and 10 for more discussions).

Weights of Items

Two principles should be borne in mind when proposing a partial credit scoring scheme. First, as discussed previously, a higher score should reflect higher underlying latent trait. Second, be aware that the maximum (highest) score on an item gives the weight of the item. If Item A has a maximum score of 4 and Item B has a maximum score of 2, then Item A has twice the weight of Item B. When the total score is computed for the instrument, you need to be aware that Item A has more weight (therefore Item A is more "important" for achieving a high score on the instrument) than Item B.

How should the weights of items be determined? Contrary to a common perception that more difficult items should receive more weight, the weight of an item should be determined by the discriminating power of the item. For example, if a partial credit item has a maximum score of four, it suggests that this item can separate people into five groups (five score groups: 0, 1, 2, 3, 4) with different and increasing latent trait. Similarly, an item with a maximum score of two can separate people into three different ability groups. While it is necessary to have increasing difficulty with increasing scores within an item, there is no reason why a score of four for Item A should be more difficult to achieve than a score of two for Item B. In determining whether a partial credit item should have a maximum score of four or two, one needs to consider whether the item has the discriminating power to divide people into more groups. One should not allocate more score points simply because the range of responses have varying degrees of accuracy, as illustrated by the above example with the computation of areas.

An example can illustrate the difference between item difficulty and item discrimination in terms of their influence on the weight (or score) of an item. Suppose a multiple-choice item is not worded well and many students are confused. As a consequence, many students randomly guess an answer, so that only about 35% of the students obtained the correct answer. The item discrimination for this item is low, since there was some random guessing. Under these circumstances, we would want to assign a low weight (score) to the item, if not deleting the item altogether. We certainly will not give this item a large weight (score) because the item is difficult (low percent-correct).

Making judgement of item discriminating power is not as easy as making judgement of item difficulty during test development. Test writers typically have good ideas about the difficulty levels of items, but it is rarely obvious to test writers how discriminating an item may be. Information on discrimination needs to come from item analysis. Scoring schemes can be adjusted based on the empirical discrimination indices obtained after conducting trial tests of items. Alternatively, using the two-parameter IRT model (2-PL) will solve the problem of having to guess the item scores a priori. In the case of the 2PL model, the item scores are estimated rather than assigned. Chapter 9 discusses the 2PL models. Chapter 3 also provides further clarification about the scoring of item responses.

Discussion Points

(1) In many cases, the clients of a project provide a pre-defined framework, containing specific test blueprints, such as the one shown in Fig. 2.4.

These frameworks and test blueprints were usually developed with no explicit consideration of the latent trait model. So when we assess items from the perspective of item response models, we often face a dilemma whether to reject an item because the item does not fit the latent trait model, but yet the item belongs to part of the blueprint specified by the clients. How do we reconcile the ideals of measurement against client demands?

(2) To what extent do we make our test "unidimensional"? Consider a spelling test. Spelling words generally have different discriminating power, as shown in the following examples.

Spelling word:	Infit MNSQ = 0.85	
(heart)		Disc = 0.82
Categories	0 [0]	1 [1]
Count	13	39
Percent (%)	25.0	75.0
Pt-Biserial	-0.82	0.82
Mean Ability	-0.08	3.63

Spelling word:	Infit MNSQ = 1.29	
(discuss)		Disc = 0.49
Categories	0 [0]	1 [1]
Count	40	42
Percent (%)	48.8	51.2
Pt-Biserial	-0.49	0.49
Mean Ability	0.76	2.40

Can we select only spelling words that have the same discriminating power to ensure we have "unidimensionality", and call that a spelling test? If we include a

Fig. 2.4 Example client specifications for a test

FINAL FORM MATRIX					
	Yr 3	Links 3/5	Yr 5	Links 5/7	Yr 7
Number	14	5	16	5	17
Space	8	2	9	2	10
Measurement	8	2	9	2	10
Chance & Data	4	2	6	2	6
Total	34	11	40	11	43

random sample of spelling words with varying discriminating power, what are the consequences in terms of the departure from the ideals of measurement?

(3) Can we assume that the developmental stages from year 1 to year 12 form one unidimensional scale? If not, how do we carry out equating across the year levels?

Exercises

In the SACMEQ project, some variables were combined to form a composite variable. For example, the following seven variables were combined to derive a composite score:

24. How often does a person other than your teacher make sure that you have done your homework?
(Please tick only one box.)

PHMWKDON

(1)	I do not get any homework
(2)	Never
(3)	Sometimes
(4)	Most of the time

25. How often does a person other than your teacher usually help you with your homework?
(Please tick only one box.)

PHMWKHLP

(1)	I do not get any homework
(2)	Never
(3)	Sometimes
(4)	Most of the time

26. How often does a person other than your teacher ask you to read to him/her?
(Please tick only one box.)

PREAD

(1)	Never
(2)	Sometimes
(3)	Most of the time

27. How often does a person other than your teacher ask you to do mathematical calculations?
(Please tick only one box.)

PCALC

(1)	Never
(2)	Sometimes
(3)	Most of the time

28. How often does a person other than your teacher ask you questions about what you have been reading?
(Please tick only one box.)

PQUESTR

(1)	Never
(2)	Sometimes
(3)	Most of the time

29. How often does a person other than your teacher ask you questions about what you have been doing in Mathematics?
(Please tick only one box.)

PQUESTM

(1)	Never
(2)	Sometimes
(3)	Most of the time

30. How often does a person other than your teacher look at the work that you have completed at school?
(Please tick only one box.)

PLOOKWK

(1)	Never
(2)	Sometimes
(3)	Most of the time

The composite score, ***ZPHINT***, is an aggregate of the above seven variables.

Q1. In the context of IRT, the value of ***ZPHINT*** can be regarded as reflecting the level of a construct, where the seven individual variables are manifest variables. In a few lines, describe what this construct may be.

Q2. For the score of the composite variable to be meaningful and interpretable in the context of IRT, what are the underlying assumptions regarding the seven indicator variables?

Q3. In evaluating the quality of test items, which one of the following is the **most un**desirable outcome for an item?

The item is difficult and less than 25% of the students obtained the correct answer
One distractor attracted only 5% of the responses. That is, one distractor is not "working" well
The percentage correct for **high** ability students is about the **same** as the percentage correct for **low** ability students
A lot of students skipped this question because they don't know the answer

Q4. In determining the maximum score of an item (e.g., an item is worth two or four marks), which of the following is the **most** important consideration?

The number of steps needed to reach the final answer
The difficulty of the question. The more difficult, the higher the maximum score should be
The range of possible responses. If there are more different responses, there should be more score points
The extent to which a question can separate good and poor students

Q5. Answer TRUE or FALSE to the following statement:

For an item where the maximum score is more than 1 (e.g., an item with a maximum score of 3), the scores (0, 1, 2, 3) should reflect increasing difficulties of the expected responses. That is, the assignment of the scores to responses should reflect an increasing ability, where a student receiving a higher score is expected to have a higher ability than a student receiving a lower score on this item.

TRUE/FALSE

Q6. Can you have a think about Questions 4 and 5. Write a short summary about the considerations of the assignment of partial credit scores **within** an item, and **across** items?

References

Adams RJ, Wu ML (2002) PISA 2000 technical report. OECD, Paris

Downing SM, Haladyna TM (eds) (2006) Handbook of test development. Lawrence Erlbaum Associates, Mahwah, NJ

Mullis I, Martin M, Ruddock G, O'Sullivan C, Preuschoff C (2009) TIMSS 2011 Assessment Frameworks. TIMSS & PIRLS International Study Center Lynch School of Education Boston College, Boston, MA

OECD (2013) PISA 2012 Assessment and analytical framework: mathematics, reading, science, problem solving and financial literacy. OECD Publishing, Paris. doi:10.1787/9789264190511-en

Osterlind SJ (2002) Constructing test items: Multiple-choice, constructed-response, performance, and other formats, 2nd edn. Kluwer Academic Publishers, New York

Reckase MD, Ackerman TA, Carlson JE (1988) Building a unidimensional test using multidimensional items. J Educ Meas 25:193–203

Further Reading

Hogan TP, Murphy G (2007) Recommendations for preparing and scoring constructed response items: what the experts say. Appl Measur Educ 20(4):427–441

Mellenbergh GJ (2011) A conceptual introduction to psychometrics: development, analysis, and application of psychological and educational tests. Eleven International Publishing, Hague, Netherlands

Netemeyer RG, Bearden WO, Sharma S (2003) Scaling procedures: issues and applications. Sage, Thousand Oaks, CA

Schmeiser CB, Welch CJ (2006) Test development. In: Brennan R (ed) Educational measurement, 4th edn. Praeger publishers, Westport, CT, pp 307–354

Wu ML (2010) Measurement, sampling and equating errors in large-scale assessments. Educ Measur: Issues Pract 29(4):15–27

Chapter 3
Test Design

Introduction

In this chapter, test design refers to the considerations for the number of items in a test, the sample size of students to take the tests, the assignment of tests to students, the arrangement of items in a test and the assignment of markers to test scripts. However, more generally, the development of the construct, framework and test blueprint discussed in Chap. 2 are all part of the test design.

The purposes of measurement can vary a great deal. Even if we focus on achievement tests, there can still be many different objectives. Test design depends greatly on the purposes of a test. For example, a test can be used for diagnostic purposes, in which case measures on individual students are the main focus. A test can be used to gather information for a state, or for a country, in which case the focus is not on individual students but on the cohort. In the case of cohort statistics, the focus may be on the percentage of students reaching a minimum standard, or the focus may be on the shape of the ability distribution and the percentages of students at different levels. For another test, the focus may be on the items for building an item bank for future uses rather than on student achievement per se. One might desire a test that can accomplish multiple purposes. But the costs and practicality of constructing and administering such tests may not be feasible. Therefore in the following we consider trade-offs and tensions between different uses of a test, and the implications of different purposes to test design.

Measuring Individuals

If the purpose of a test is diagnostic for individual students, then the construct should not be too broad. Since a broad construct calls for many items, and it may not be practical for a student to sit a very long test.

M. Wu et al., *Educational Measurement for Applied Researchers*,
DOI 10.1007/978-981-10-3302-5_3

Magnitude of Measurement Error for Individual Students

To gauge whether a test provides sufficient accuracy for measuring individuals, there are two considerations. First, we need to know the accuracy of the ability estimates. Second, we need to know what accuracy is sufficient for the purposes of the test. To answer the first question, the computation of measurement error will inform us of the accuracies of ability estimates. The notion of measurement error can be explained as the possible variation in a student's test scores if similar tests are administered. The variation in scores is due to the fact that each test only samples a small set of a student's capabilities reflecting the construct. That is, when a test is administered, there is some uncertainty associated with the test score, not because the test contains errors but because a test provides limited information about a student's ability in the domain being tested. Should different test developers write the test (to the same test construct and test blueprint), a student's score will likely be different.

It is unfortunate that the word "error" leads some to think that measurement error is caused by errors in test items or errors made in processing test result. Actually, the quality of test items is not reflected in the computation of measurement error per se. It is reflected in the test reliability and validity. To evaluate whether measurement error is large, we need to look at the relative magnitude of the measurement error compared to the overall spread of the ability distribution, in other words, effect size.

Measurement error is directly related to the test length. A student's test scores on similar tests of 5-items will likely vary a great deal. A student's test scores on 100-item tests will not vary as much. To get an idea about the magnitude of measurement error, Fig. 3.1 shows the ability distribution estimated using a state-wide mathematics test for grade 5 students. Measurement error for this test can be calculated. We can also predict the measurement error for similar tests with varying number of items. The 95% confidence intervals of a student's ability measure in relation to the spread of the ability distribution are shown in Fig. 3.1 when such a test has 30 items and 60 items. The ability measure is in the logit unit (see Chap. 7 for the meaning of logit). The derivation of the confidence intervals is given in Appendix 1.

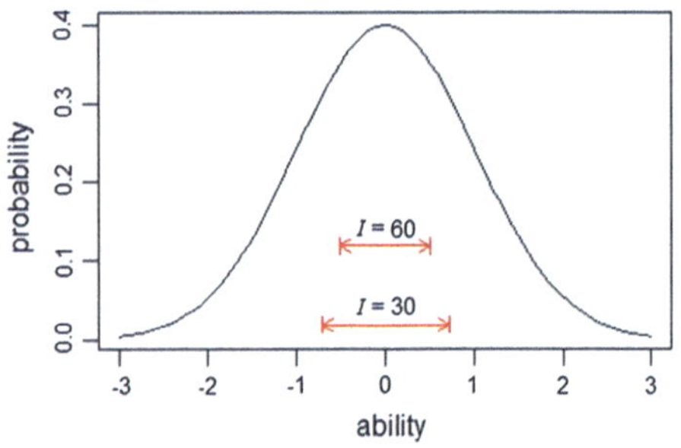

Fig. 3.1 Ability distribution and 95% confidence intervals for ability measures

Figure 3.1 shows the ability distribution which has been standardised to have a mean of 0 and standard deviation of 1. The two (red) horizontal lines with arrows show the width of the 95% confidence intervals for an individual student's ability estimate, as a function of test length (value of I). For example, reading from the graph, if the test length is 30, a student's ability estimate is likely to vary between −0.7 and 0.7 should the student sit similar tests of 30 items. Given that the student only sat one test, one needs to be aware that the ability estimates obtained from similar tests could vary in a range of 1.4 (i.e., from −0.7 to 0.7). As the test length increases, the width of the confidence interval becomes smaller, indicating more accuracy in the ability estimate. However, even for tests of 60 items, the 95% confidence interval is still quite wide (reading from the graph, about 1.0 in length). Appendix 1 provides a table of the magnitude of the measurement errors.

While every test is a little different in terms of the relative size of the measurement error to the standard deviation of the ability distribution, our experience shows that for standardised tests typically used in large-scale surveys, the order of magnitude of measurement error is similar to those shown in Fig. 3.1. As a rough guide, a lower bound for measurement error in logit unit is given by $\sqrt{\frac{4}{I}}$ where I is the number of items (actually, score points) in a test (see Appendix 1) and the 95% confidence interval for an ability estimate is $\pm 2 \times \sqrt{\frac{4}{I}}$. If the 95% confidence interval needs to have a width of less than 0.5, say, the number of items, I, needs to be more than 256 (in the equation $0.5 = \left(2 \times \sqrt{\frac{4}{I}}\right) - \left(-2 \times \sqrt{\frac{4}{I}}\right)$, solve for I). It may take up to 6 h for a student to work through 256 items if it takes 1 h to complete 40 items. This does not sound feasible in practice. Even then, a confidence interval of 0.5 may still be too large for some purposes. This leads to the next consideration of determining what accuracy is sufficient. But before discussing about how to decide on adequate accuracy, we will first introduce the notion of standardised units where scores are expressed in standard deviation units to provide a convenient way for making comparisons of quantities when there are different units of measurement.

Scores in Standard Deviation Unit

To compare scores from different assessments where the scale factor and the origin of a measurement scale are arbitrarily set, we will compute statistics in units of the standard deviation of the distribution of interest, in much the same way as the unit of effect size. For example, for a 30-item test, the measurement error is around 0.37 in the above example. This corresponds to 0.37 standard deviation units, if the ability distribution has a standard deviation of 1. In the example below (Fig. 3.2), student achievement scores have been scaled with a range between 200 and 800. While this scale is not directly comparable with the scale in Fig. 3.1, scores in standard deviation units can be compared.

What Accuracy Is Sufficient?

To determine whether an achieved accuracy is sufficient, we need to assess whether the aims of the measurement can be met. For example, if we only want to divide students into three ability groups with roughly the same range, a 30-item test will provide sufficient precision for that, as shown in Fig. 3.1. But at the same time, teacher judgements probably can provide us that information with sufficient accuracy already. Suppose the purpose of our test is to monitor student growth. We need to first make a guesstimate of the magnitude of the growth before we can design instruments to measure that growth with sufficient accuracy. As an analogy, if we want to measure the amount of baking powder for a cake, we need accuracies to the nearest gram and we will not use a scale that measures in whole kilograms. To estimate growth measures, we use some empirical data obtained in a national assessment of numeracy of students in Australia (NAPLAN 2010). Figure 3.2 shows the ability distributions from NAPLAN tests for Grades 3, 5, 7 and 9 (grey rectangular boxes showing 95% of each distribution), and the average achievement at each grade (squares at the centre of the distributions). From this graph, we can estimate a relative growth rate in relation to the spread of each distribution. The standard deviation of each distribution is around 70 NAPLAN score points. The growth rate per year is larger in lower grades than in higher grades. Overall, there is an increase of 190 NAPLAN score points over 6 years (from Grade 3 to Grade 9), so the average growth rate per year is 32 points (190 points divided by 6 years). The growth rate can be expressed in standard deviation unit. In this case, the growth rate is about 0.46 standard deviation unit (32 divided by 70). In education literature, yearly growth rate has typically been found to be about 0.5 of a standard deviation of the ability distribution (e.g., Thissen and Steinberg 1997).

Expressed in standard deviation units, given that the average yearly growth rate is around 0.5, and the measurement error for a 30-item test is around 0.37 (with 95% confidence interval of the growth measure around 1.4), it is not expected that two 30-item tests administered one year apart can provide the accuracy to measure an individual student's growth with sufficient precision. This is similar to using a scale calibrated with kilograms to measure a few grams of baking powder. We need more accurate instruments to measure individual growth. To reduce the 95% confidence interval of growth measures so that growth (around 0.5 standard deviation unit) can be measured with reasonable accuracy, we need many hundreds of items!

In general, the technique used to determine sample sizes (whether the number of items or the number of students) begins with an estimation of the order of magnitude of the statistic to be measured, and then an estimation of the accuracies that can be achieved with various sample sizes. Appropriate sample sizes can be chosen to meet the measurement purposes. In general, there is no one single recommendation regarding adequate sample size. Each purpose of measurement has its own requirement regarding sample size.

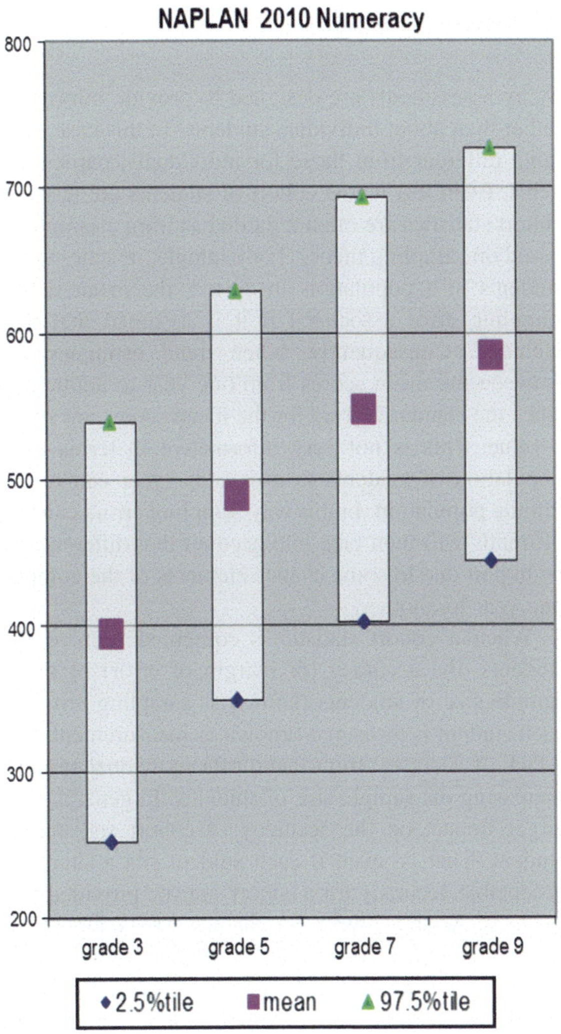

Fig. 3.2 Student numeracy scores distributions (NAPLAN 2010)

Summary About Measuring Individuals

The main message is that a one-off test does not provide very accurate information at the individual student level other than an indicative level of whether a student is below average, at average or above average. This lack of accuracy should not surprise us. If ever one single test of 30 items is used for high-stakes purposes such as selection into colleges or awarding certificates, we should be very wary of the results. This message is particularly relevant in the current climate of the proliferation of standardised tests. Results from these tests have often been over-interpreted.

Measuring Populations

Many assessments are designed to provide information about a cohort of students rather than about individual students. In this case, considerations for test design are quite different from those for individuals, particularly about the sampling of students. Even if a whole cohort of students are tested rather than a selected sample, cohort statistics are often regarded as from a sample, and many inferences are made based on sampling theory. For example, in state-wide or nation-wide tests where all students in a population are tested, the mean score for the cohort will have no sampling error associated if it is assumed that the whole population has been included. Consequently, when trend estimates are computed, any difference between the mean scores from one year to another will be statistically significant, since the standard errors for the mean scores are zero, leading to an infinitely small p-value. This is not very informative in terms of reporting trends. Instead, the population of students in a calendar year can be regarded as a sample from an infinite population. In this way sampling errors can still be computed. An evaluation of trends will then take into account that differences between the mean scores can be in part due to some chance elements of the composition of the populations from one year to another.

When a cohort statistic is computed such as the mean score of a group of students, the accuracy (or margin of error) of the statistic depends on both the sample size of students (known as sampling error), as well as on how accurately each student is measured (known as measurement error). For the group mean score, a lack of accuracy from measuring individual students can be compensated for by increasing the sample size of students. In general, the sample size of students has a larger impact on the accuracy of cohort statistics than the test length for each student. That is, even if each student sits a short test, we are still able to obtain acceptable accuracy for a cohort statistic provided there are enough students taking the tests. As an example, we can ask 5000 people their age groups (e.g., age groups in the range of 0–10, 11–20, 21–30 etc.) so the information on individual person's age is not accurate. But if we average the age groups of 5000 people, we can still obtain an average age that will not be greatly different from the average of everyone's actual age.

In the case of state-wide testing, one common purpose is for monitoring student standards in a subject domain such as mathematics or reading. In this case, the constructs are broad and the tests need to have a large number of items to cover the subject domain. This relates to test validity issues. Since it is impractical to administer a very long test to each student, rotated test booklets are often designed where each test booklet contains only a small number of items. The test booklets are randomly distributed to students. In this way, the subject domain is covered by many items, but at the same time there is not too much burden for each student to take the tests. The lack of accuracy for individual student's ability measure is compensated by the large number of students taking the tests. The content validity of the assessment is supported by the inclusiveness of many items in the assessment.

Computation of Sampling Error

The computation of sampling error depends on the sampling design (i.e., how the sample of students is selected). In the simplest case, if students are randomly selected, the standard error for the group mean for a statistic X is given by $\frac{\sigma_X}{\sqrt{n}}$, where σ_X is the standard deviation of X and n is the sample size. For example, if we know that the standard deviation σ_X is 1 and we want the standard error to be less than 0.05, the sample size required will be 400 or more. However, in many large-scale assessments, the sampling method is not simple random sampling. Instead, cluster sampling method is used (schools are sampled first and then students are sampled within sampled schools). There are at least two reasons for this. First, for most countries, if 400 students in a population (e.g., Grade 8 students) are randomly selected, the 400 students are likely to come from 400 different schools. This could potentially increase test administration costs if 400 schools need to participate in the assessment. Second, any analysis at the school level will not likely have sufficient data if only one student is selected from each school. Consequently, for large-scale assessment programs, typically, around 150 schools are selected, with one class or 30 students being selected from each sampled school. However, such cluster sampling reduces the "efficiencies" of the samples in that the standard errors of sample statistics are typically much larger than the standard errors of simple random samples. Apart from cluster sampling, the sampling design may include stratification of the population and other sampling considerations. In these cases, the computation of the sampling error becomes complex. These topics are beyond the scope of this book. Interested readers can refer to technical reports for OECD's PISA and IEA's TIMSS for some examples of complex sampling designs and methods for computing sampling errors (e.g., OECD 2012; Olson et al. 2008).

Summary About Measuring Populations

For large-scale assessments focusing on the measurement of populations, there need to be many items to cover a subject domain. Rotated test booklets can be distributed to students at random, so each student can take just a subset of the items. Typically, a sample of students from the population can be selected to take the tests. Provided the sample size is sufficient and the sample design is sound, it is not necessary to test every student in the population. However, complex sampling requires advanced techniques in the computation of statistics and their standard errors.

Placement of Items in a Test

When arranging items in a test, it is often recommended that a test should begin with easier items so as to alleviate any anxiety the test takers may have. This makes a great deal of sense. In this section, we will discuss the impact of test taker fatigue on item difficulties, using the PISA data as an example. (See http://www.oecd.org/pisa/pisaproducts/pisa2003/.) PISA 2003 has 13 rotated test booklets. For each test booklet, there are four blocks of test items placed in each booklet. We refer to the four positions of item sets as Block 1, Block 2, Block 3 and Block 4 in order of their placement in a test booklet. In PISA 2003, each mathematics item appears in four test booklets in different Block positions. That is, each mathematics item appears once in the first position of a booklet, once in the second position, once in the third position and once in the fourth position of some other test booklets. For example, Table 3.1 shows the percentages correct of the first five items in the OECD database when the items appear in four different positions.

It can be seen that the percentages correct tend to decrease as the item is placed in the latter part of a booklet (i.e., numbers in each column of Table 3.1 are generally decreasing). More notably is the large decrease in percentage correct when an item is placed in the last part (Position 4) of a test booklet. Figure 3.3 shows the percentages correct graphically. It is quite clear that there is a downward trend as an item is placed in the latter parts of a test booklet. While it is not clear whether this is due to the lack of motivation or mental and physical fatigue, we will term this effect "fatigue effect" for convenience. The difference in percentages correct between Position 1 and Position 4 range between 5 to 15% for the example items given. The five items were not specially chosen with decreasing percentages correct. They were the first five items in the OECD database. Assuming this provides an indicative range of fatigue effect for the whole assessment, the effect is considerably large.

Implications of Fatigue Effect

What Fig. 3.3 shows is that the estimation of item difficulties will be influenced by the position at which an item is placed in a test booklet. The estimated item

Table 3.1 Percentages correct of five mathematics items in four different positions in test booklets

	Percentage correct				
	View room	Bricks	Walking	Cube painting	Growing up
Position 1	72.9	43.9	38.1	61.6	65.2
Position 2	75.6	41.1	37.1	60.6	62.8
Position 3	70.8	35.8	33.1	59.9	60.2
Position 4	67.2	30.1	30.2	49.3	50.4

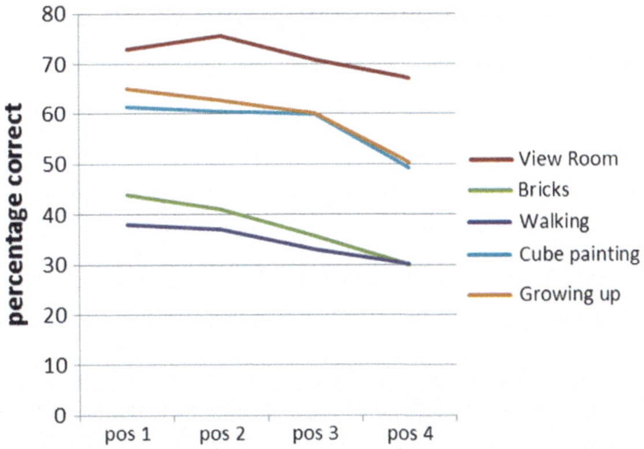

Fig. 3.3 Percentages correct of five items in four positions of a test

difficulty for an item may be quite different if the item appears in another test, depending on the item position. This has an important implication for equating tests (see Chap. 12), as the assumption of item invariance needs to be made when equating is carried out with link items that appear in different tests. For example, suppose item X appears in both Test A and Test B and is one of the set of link items used for equating Test A and Test B. Item X appears at the beginning of Test A and at the end of Test B. If we use the estimated item difficulty for Item X obtained from Test A data as the item difficulty for Item X in Test B, we under-estimate the item difficulty in Test B, leading to an under-estimation of Test B students' abilities.

While this is a complex problem to solve, one way to lessen the impact of fatigue effect on item difficulty estimation is to have a rotated test booklet design that is balanced in the sense that an item appears in several different positions in different test booklets, so that the average item difficulty for the item is computed across the difficulties at different positions.

Balanced Incomplete Block (BIB) Booklet Design

As an example of a balanced incomplete block booklet design, Table 3.2 shows a 7-booklet test design with 7 clusters (C1 to C7) of items placed in three blocks of each test booklet. The term "cluster" is used to denote the division of all assessment items into groups. In this example, there are around 100 items in total. The 100 items are divided into 7 clusters, with each cluster containing around 14–15 items. Since three clusters are placed in each booklet in the three block positions, each booklet contains around 40–45 items.

The seven clusters, C1 to C7, are placed in one of three positions (Blocks 1, 2 and 3) in seven test booklets. The design shown in Table 3.2 has the following characteristics.

Table 3.2 7-booklet BIB design

Booklet	Block 1	Block 2	Block 3
1	C1	C2	C4
2	C2	C3	C5
3	C3	C4	C6
4	C4	C5	C7
5	C5	C6	C1
6	C6	C7	C2
7	C7	C1	C3

1. Each cluster appears in each position (Block 1, 2 or 3) once. For example, C1 appears in position 1 in booklet 1, position 2 in booklet 7 and position 3 in booklet 5.
2. Each possible pair of clusters appears together in a booklet once. For example, Clusters 4 and 6 appear together in booklet 3. Clusters 3 and 7 appear in booklet 7.

The above two characteristics show a "balance" in the design. The design is incomplete in the sense that each booklet does not contain all clusters. Using a BIB test design greatly enhances the strengths of the links between the test booklets. It also moderates the impact of fatigue effect as each item appears in different positions of the test booklets.

There are other examples of BIB designs. In PISA 2003, a 13-booklet BIB design was used, as shown in Table 3.3 (OECD 2005), showing the placement of mathematics, science, reading and problem solving item clusters.

Whether a 7-booklet, 13-booklet or other BIB test design should be used depends on the total number of test items in the assessment and the amount of time each test taker can take a test. The test length of each cluster and the number of blocks in a booklet determine the total test length for each test taker. For example,

Table 3.3 PISA 2003 test design

Booklet	Block 1	Block 2	Block 3	Block 4
1	M1	M2	M4	R1
2	M2	M3	M5	R2
3	M3	M4	M6	PS1
4	M4	M5	M7	PS2
5	M5	M6	S1	M1
6	M6	M7	S2	M2
7	M7	S1	R1	M3
8	S1	S2	R2	M4
9	S2	R1	PS1	M5
10	R1	R2	PS2	M6
11	R2	PS1	M1	M7
12	PS1	PS2	M2	S1
13	PS2	M1	M3	S2

under the 13-booklet design, if each cluster contains 20 items, then each booklet will have 80 items (4 blocks per booklet). This may be too many for a student to take. On the other hand, the total number of clusters and the length of each cluster determine the total amount of materials that can be tested. For example, if there are 13 clusters and each cluster has 20 items, then the total amount of testing material is 260 items (13 × 20). Consequently, the test design needs to take into account the amount of testing materials in total and the testing duration for each student.

Arranging Markers

For open-ended items such as essays, markers are often required to make judgements on students' responses. Frequently, steps are taken to minimise marker differences. For example, control scripts (e.g., a set of essays used for marker training and monitoring) are provided so marker variation can be assessed. On-going monitoring of marker harshness/leniency is also a good quality control practice. The following is an example of marker differences even when experienced and well-trained markers were used for marking essays. The data set is from a state-wide assessment on essay writing. Twenty markers participated in the marking of essays. Each essay was marked by two markers. The maximum score for each essay is 7 score points. Figure 3.4 shows the results of an IRT analysis that computes the expected scores of each of the 20 markers as a function of student ability (see Chap. 13 for details on IRT analysis of markers). For example, for Marker 1, the average (expected) score for students with an ability of 2 is 4 score points. Each curve in Fig. 3.4 shows the expected scores curve of one marker.

The curves are increasing with ability (theta), indicating that higher ability students have higher expected scores. However, the band of curves has a width of about one score point. That is, the most lenient marker (curve on the top of the band) and the most harsh marker (curve at the bottom of the band) differ by about one score point out of seven. This is quite a large difference (more than 10% of the total mark), even after the markers had training to improve marker consistency.

Carrying out an analysis such as the one shown in Fig. 3.4 has at least two purposes. First, markers can obtain feedback on their leniency/harshness for future improvement. Second, an IRT analysis with explicit estimations of rater harshness (see Chap. 13 on Facets analysis) can make adjustments to students' ability estimates taking into account of marker harshness. While Chap. 13 discusses the details of such analyses, in this chapter, we will explain about how to design a marking scheme so that marker effects can be estimated. Clearly, to reduce marker effect, each essay could be marked by all markers. But such a marking scheme is likely to be very costly. Marker effects can be estimated without all markers marking all essays. The following is an example.

Figure 3.5 shows an excerpt of the data that underlie the results shown in Fig. 3.4.

Fig. 3.4 IRT expected scores curves for 20 markers

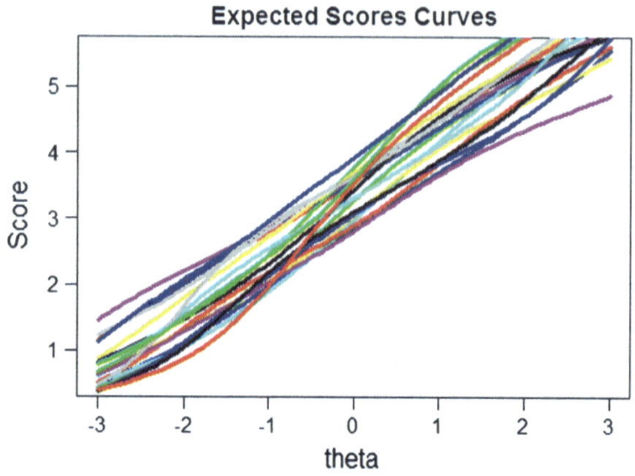

Fig. 3.5 Excerpt of a data file of essay marks

11239	66532	2	5
11239	92060	4	4
92060	31256	3	5
01884	25181	1	1
25181	31256	2	0
66532	66700	2	1

....

Each row in Fig. 3.5 shows the results for one student. Each student's essay has been marked by two markers. The first two columns show marker IDs; the third column shows the mark given by the first marker; and the fourth column shows the mark given by the second marker. For example, Marker 11239 gave a score of 2 to Student 1, while Marker 66532 gave a score of 5. From this one data record, one might make an initial guess that Marker 11239 is relatively harsher than Marker 66532. For the second student, Marker 11239 gave a score of 4, so did Marker 92060. So both markers seem to have the same leniency. We might then infer that Maker 66532 is more lenient than Marker 92060, even though these two markers have not marked any essay in common. Provided there are linkages across all markers, whether through direct link from marking the same essay, or from indirect links through other markers, we will be able to compare all markers in terms of their leniency/harshness. On the other hand, if there are two groups of markers, and markers in Group A have not marked any essays in common with markers in Group B, then we will not be able to compare these two groups of markers. If, on average, Group A markers gave higher scores than Group B markers, we will not be able to conclude whether this is because Group A markers are lenient, or the students are of higher ability, since Group A marked different students from Group B. That is, student ability and marker harshness are completely confounded.

In summary, to be able to carry out marker harshness comparisons, there must be links across markers when we distribute essays for marking. In the case where every essay is only marked by one marker, it is generally difficult to compare marker harshness.

Summary

This chapter discusses about the sample size of items, sample size of students, the placement of items in a test and assignment of markers to test scripts. Sample sizes need to be sufficient to provide the accuracy required for the purposes of an assessment. In general, the accuracy of a statistic depends on the number of pieces of information we have about that statistic. For example, the accuracy of item statistics depends on the number of students taking each item. The accuracy of student measures depends on the number of items each student takes. The accuracy of marker harshness estimates depends on the number of test scripts a marker marks. For group statistics, the sample size of students has more impact on the accuracy of measures than the number of items. If statistics for sub-groups are required (e.g., for a state or for a socio-economic group), then the sample sizes for the sub-groups are also important considerations. We often compute the required sample size for a whole cohort and forget that there needs to be sufficiently large samples for statistics for sub-groups.

In designing tests, whenever possible, apply the principles of "balance" and "randomisation": a balanced range of item difficulties in a test; a balanced item rotation in test booklets, a balanced assignment of markers to essays, etc. When real-life situations have "unbalanced" characteristics, such as differences between schools and classes, we can distribute rotated test papers at random within each class so as to avoid having high performing schools receiving particular test booklets. Randomisation at the lowest sampling level is typically a good recommendation.

Finally, we would like to say a word about "fairness". One common misperception is that students can only be compared if they take the same test. Consider an example. If we were able to administer all possible Grade 5 mathematics test items to a student, the student can answer 60% of the items. When 40 items are placed in a test, say, the student may not obtain exactly 60% (24 items) correct. The student's score depends on the particular set of items in the test, as well as his/her performance on the day of the test. Should a different set of test items be selected, equally representing Grade 5 mathematics, the student's test score could be different, not because there are errors in the test or in the data processing, but because by chance the student may know more or less of the content of a particular test. This is the notion of measurement error. A student's score on a 40-item test typical could vary within a 10 score point range. For a given test, Student A could obtain 20 out of 40. But on another similar test, the same student could obtain 30 out of 40. Consequently, some tests may "favour" Student A, and some tests may disadvantage Student A, so that when every student takes the same test, the test is not equally "fair" to all students, because a student could do better (or worse) on another similar test. If a set of similar tests are randomly distributed to students, and students take different tests, it is no more unfair than with all students taking the same test. For this reason, when rotated test booklets are used in an assessment, it is no more unfair than having a single test for all students.

The issue with administering the same test to students is that there will be a lack of curriculum coverage. This leads to validity issues as well as problems with equating tests from year to year because of content differences. In some state-wide testing programs, the administration of a single test for everyone has greatly limited the usefulness of the assessment, just because test administrators believe that it is only fair when all students take the same test.

Discussion Points

1. Stakeholders of assessments have competing demands. Discuss the information typically desired by students, parents, teachers and education authorities, the tensions between them, and how different information can be obtained through assessments.

Exercises

Q1. To increase the precision of each student's ability estimate, which one of the following is the most important factor?

Increase the test length (or the score points on a test)
Increase the sample size of students taking the test
Make sure there are no errors in scoring the items
Make sure there are no errors in the test questions

Q2. To increase the precision of cohort statistics, which one of the following is the most important factor?

Increase the test length (or the score points on a test)
Increase the sample size of students taking the test
Make sure there are no errors in scoring the items
Make sure there are no errors in the test questions

Q3. If there are 100 items in a test (or, the maximum score is 100), estimate the 95% confidence interval of an ability estimate (use Eq. (3.2) in the Appendix). What is the effect size of the confidence interval if the ability distribution has a standard deviation of 1?

Q4. If students are randomly sampled from a population where the ability distribution has a standard deviation of 1, what is the sampling error of the mean ability for a sample of 1000 students?

Q5. Ignoring measurement error, what sample size for a simple random sample of students will lead to a 95% confidence interval of ±0.01 for the mean ability, if the ability distribution has a standard deviation of 1?

Q6. Design a BIB test booklet rotation scheme for 3 clusters of items. How many blocks and how many test booklets are needed?
If each student can take 60 min of test materials, what is the total amount of materials (in terms of test minutes) that can be included in your test design?

Q7. Discuss the following test design in terms of balance and linkages of clusters.

Booklet	Block 1	Block 2
1	C1	C2
2	C2	C3
3	C3	C4
4	C4	C5
5	C5	C6
6	C6	C7
7	C7	C1

Q8. Four markers (A, B, C, D) will mark 6 students' essays. Each student's essay will be marked by two markers. The following shows a design of assigning students' essays to markers.

Student	First marker	Second marker
1	A	C
2	A	C
3	A	C
4	B	D
5	B	D
6	B	D

Discuss any advantages or disadvantages of the above marking scheme. Can you propose any alternative marking scheme that may work better? Give reasons for why your design may be better.

Q9. In real-life, it is often too costly to mark each student's essay more than once. If every essay is marked only once, what assumptions must be made if the average mark given by each marker is used to reflect the marker's harshness/leniency?

Q10. Given that a group of markers have different harshness/leniency, if test scripts are *randomly* distributed to markers, is this "fair" for each student in the sense that there is no marker bias for individual students? What marking design can reduce marker bias to make it fairer for individual students?

Appendix 1: Computation of Measurement Error

Suppose a test has I dichotomous items with item difficulties $\delta_1, \delta_2, \ldots, \delta_I$. Let θ_n denote the ability of the nth examinee, and $\hat{\theta}_n$ denote the maximum likelihood estimate of θ_n. Then, it can be shown that the measurement error is equal to $\sqrt{\operatorname{var}\left(\hat{\theta}_n\right)}$, where

$$
\begin{aligned}
\operatorname{var}\left(\hat{\theta}_n\right) &= -\left[\frac{\partial^2 \lambda(\Theta|\mathbf{X})}{\partial \theta_n^2}\right]^{-1} \\
&= \left[\sum_{i=1}^{I} \Pr\left(X_{ni} = 1; \hat{\theta}_n, \hat{\delta}_i\right)\left(1 - \Pr\left(X_{ni} = 1; \hat{\theta}_n, \hat{\delta}_i\right)\right)\right]^{-1} \\
&= T^{-1}
\end{aligned}
\tag{3.1}
$$

for dichotomous items, where T is the test information function and $\lambda(\Theta|\mathbf{X})$ is the log likelihood function of the item responses.

If we assume that all items have a difficulty value of 0 on the logit scale, and the ability of a student is also 0 (i.e., well-targeted test), then $\Pr\left(X_{ni} = 1; \hat{\theta}_n, \hat{\delta}_i\right)$ is $\frac{1}{2}$ in Eq. (3.1) in the case of the Rasch model, so that the measurement error is

$$
\sqrt{\left(\sum_{1}^{I}\left(\frac{1}{2} \times \frac{1}{2}\right)\right)^{-1}} = \sqrt{\frac{4}{I}}
\tag{3.2}
$$

where I is the number of items. Equation (3.2) is useful for obtaining the order of magnitude of measurement error. It provides a lower bound for measurement error as we have assumed that all item difficulties matched the student's ability. When items do not match a student's ability (e.g., items are too difficult or too easy for a

Table 3.4 Test length and measurement error

Test length (max score on test)	Measurement error (lower bound)	Width of 95% confidence interval
30	0.37	±0.72 = 1.43
40	0.32	±0.62 = 1.24
50	0.28	±0.55 = 1.11
60	0.26	±0.51 = 1.01

student), the measurement error will be large than that estimated by Eq. (3.2). Table 3.4 shows the magnitude of measurement error as a function of test length.

References

NAPLAN (2010) National Assessment Program—literacy and numeracy. National report. Retrieved 7 Dec 2012, from http://www.nap.edu.au/verve/_resources/NAPLAN_2010_National_Report.pdf

OECD (2005) PISA 2003 technical report. PISA. OECD Publishing. Retrieved 7 Dec 2012 from http://www.oecd.org/edu/preschoolandschool/programmeforinternationalstudentassessmentpisa/35188570.pdf

OECD (2012) PISA 2009 technical report. PISA. OECD Publishing. Retrieved 26 May 2013 from http://dx.doi.org/10.1787/9789264167872-en

Olson JF, Martin MO, Mullis IVS (2008) TIMSS 2007 technical report. TIMSS and PIRLS International Study Center, Lynch School of Education, Boston College. Retrieved 7 Dec 2012, from http://timss.bc.edu/timss2007/PDF/TIMSS2007_TechnicalReport.pdf

Thissen D, Steinberg L (1997) A response model for multiple-choice items. In: van der Linden WJ, Hambleton RK (eds) Handbook of modern item response theory. Springer, New York

Further Reading

van den Linden WJ (2005) Linear models for optimal test designs. Springer, New York

Wendler CL, Walker ME (2006) Practical issues in designing and maintaining multiple test forms in large-scale programs. In: Downing SM, Haladyna TM (eds) Handbook of test development. Lawrence Erlbaum Associates, Mahwah, pp 445–469

Chapter 4
Test Administration and Data Preparation

Introduction

This chapter highlights some steps in test administration and the preparation of data for test analysis, including data collection, coding and data cleaning. The key to the success of test administration is careful planning and management. From printing the test booklets to conducting the test, every step needs to be closely managed and nothing should be left to chance. For example, there could be security issues related to test papers, and attendance issues related to participating students. The whole process calls for competent management skills.

The key to the success of data processing is the development of an automated procedure that retains flexibility in editing and recoding item response data. We frequently need to analyse a dataset repeatedly because of errors in the data, change of scoring rules, deletion of poorly performing items, and many other reasons. It is essential to develop a set of computer programs to carry out recoding and data cleaning tasks. One should never make changes to the data in a manual way, as it is not only time-consuming but also error-prone.

The planning of the collection of data should begin as early as possible, in fact, at the conception of the assessment. Data collection and preparation should be an integral part of the assessment design, and it should not be an after-thought when the tests have already been designed or even after the tests have been administered.

Sampling and Test Administration

If the sample of respondents is a convenience sample and the test is a one-off test, there perhaps isn't a great deal to plan for test administration. However, there are limitations to the use of such tests, as results cannot be generalised to a population and be more

M. Wu et al., *Educational Measurement for Applied Researchers*,
DOI 10.1007/978-981-10-3302-5_4

widely used. Consequently, many assessment programs test a representative sample of a defined population, or test a whole cohort of students of a population.

Sampling

While sampling is not discussed in depth in this book, in this section we outline the key elements of sampling in educational assessment contexts for collecting samples of students to represent a population. Chapter 3 briefly introduces the concept of cluster sampling. In the following, we assume that cluster sampling is used where schools are first selected and then students are sampled from selected schools. A list of steps in sampling is given below.

1. Clearly define the target population of interest. For example, this could be grade 8 students in schools in a region, or year 13 students in schools. Without a clear population definition, sampling cannot be done to represent the population.
2. Identify all schools in which students in the target population are enrolled. Make a list of these schools. This list is called a sampling frame. This school sampling frame will typically contain school name, school information (e.g., address, school type, geolocation) and the enrolment size for each grade in each school. Figure 4.1 shows an example school sampling frame.
3. Decide on the degree of accuracy of the results of interests so that sample sizes can be computed. With a two-stage sampling design where schools are selected first and then students are sampled from the selected schools, the standard errors of mean scores are typically much larger than those from simple random sampling. The magnitude of the standard error depends on the *design effect*. The design effect is the factor by which the sample size of a simple random sample needs to be inflated for a two-stage sampling design, to achieve the same accuracy for the statistic of interest as for a simple random sample. If a country has been a participant in international studies such as TIMSS, PIRLS or PISA, you can get an estimate of the design effect from these studies. Typically, you can expect the sample size required for a two-stage sampling design to be 3–8 times larger than the sample size for simple random samples, if 30–35 students are sampled from each selected school. The design effects due to complex

School ID	School Name	School address	State	School type	School location	Grd 1 enrol	Grd 2 enrol	Grd 3 enrol
1001	St Les ..	…	Pe	catholic	urban	68	74	70
1036	Rowan ..	…	Un	govn't	rural	18	17	16
…	…							

Fig. 4.1 An example school sampling frame

Table 4.1 Design effect for PISA 2003 Mathematics (OECD 2009)

Country	Design effect	Country	Design effect	Country	Design effect
Australia	6.25	Hungary	4.19	Norway	2.68
Austria	5.52	Iceland	0.77	Poland	3.25
Belgium	3.75	Ireland	3.08	Portugal	6.94
Canada	11.67	Italy	11.24	Slovak Republic	3.79
Czech Republic	8.42	Japan	7.42	Spain	7.64
Denmark	3.57	Korea	6.47	Sweden	3.31
Finland	2.63	Luxembourg	0.43	Switzerland	9.68
France	3.09	Mexico	53.89	Turkey	13.33
Germany	4.86	Netherlands	4.48	United Kingdom	6.34
Greece	7.89	New Zealand	2.17	United States	4.87

sampling for PISA 2003 Mathematics for 15 year-olds are shown in Table 4.1 (OECD 2009).

4. Use a probability sampling method to select a sample of schools. Probability sampling means that every school and student in the target population has a positive probability of being selected, and these probabilities can be computed. In the simplest case, simple random sampling can be used. To improve sampling efficiency (i.e., to reduce sampling error), stratified and systematic sampling can also be used. Stratification typically refers to grouping the schools in the target population into strata, such as by geographical location or by school types (e.g., public, private) to ensure that each stratum has a representative sample of schools. Standard errors can be reduced by using stratification, provided that the stratification variables are related to the performance being measured. For example, a sampling frame may be stratified by geolocation: urban, rural and remote. If, on average, students in urban regions tend to perform better than students in rural areas, and a great deal better than students in remote regions, then, stratifying the sampling frame into geolocations will help us achieve a more representative sample, since we ensure that schools in all three geolocations are proportionally selected. If we leave it to chance by using simple random sample, we may not necessarily have a sample that reflects the proportions of schools in each geolocation.
5. For the selected schools, construct lists of students (sampling frames of students) in these schools. Note that in step 2, the sampling frame is a list of schools only, not individual students. This is because it is often too difficult to compile the list of students in the whole target population. Compiling the list of students in sampled schools (instead of all schools) can significantly reduce the amount of

work. To find eligible students (i.e., students satisfying the target population definition) and their background information (e.g., name, gender, age, class), the selected schools will often need to be contacted, if detailed enrolment data is not available through other sources.

6. Use a probability sampling method to sample students in selected schools. This can be just simple random sampling, or cluster sampling by selecting intact classes, or by selecting classes first and then students in the selected classes. Typically in international large-scale studies, around 30–35 students are selected per school, and around 150–180 schools are selected per country, to give an accuracy of the equivalent of a simple random sample of 400 students.
7. Compute sampling weights. In the above two-stage sampling process (select schools first and then students), it is important that the probabilities of school selection and student selection are clearly computed from the sampling design. Knowing the probability of selection will enable us to compute sampling weights. For example, if a school has a probability of 0.1 of being selected (i.e., one school is selected out of 10), then this school should represent 10 schools (inverse of the probability of selection, 1/0.1). Therefore the school sampling weight is 10. Similarly, the inverse of the probability of selecting a student is the sampling weight of the student. For example, if there are 100 eligible students in a school and 30 are selected, the student weight is 100/30, being the inverse of the probability of selection. That is, each student represents 3.3 students in the school for the target population. The final student weight is the product of the school weight and the student weight. In the example, the final sampling weight for a student is $10 \times \frac{100}{30} = 33.3$. One can think of sampling weight as the number of students in the target population represented by a sampled student.
8. In all analysis of student results, it is essential to use sampling weights to weigh each student's result so that any aggregation of results can reflect the characteristics of the population.

For a more detailed description of sampling procedures and the computation of sampling weights, see PISA and TIMSS technical reports (e.g., OECD 2009; IEA 2008).

Field Operations

In some assessments, a pilot test for checking the quality of the test items is first conducted before a main study is carried out. For pilot tests, there may not need to be stringent sampling procedures. However, for the main study, the sampling procedures and test administration process needs to be carefully executed so that the data collected can reflect the population performance. The following is a list of key steps in field operations of test administration.

1. Schools must be contacted about test administration dates so that required actions can be arranged with ample amount of time. For example, students and parents may need to be informed and permissions need to be obtained. Test booklets should be packed with lists of sampled students and sent to schools prior to test administration. If rotated test booklets are used, assign the booklets to sampled students when packing the test booklets. Do not leave it to the test administrators for the assignment of test booklets. The test booklets should be randomised within a class (if classes are selected). Typically a random booklet number is drawn (say, 6) and then the booklets are assigned sequentially to students in a list (e.g., 6, 7, 8, …, 1, 2, …). If the booklets are distributed starting from booklet 1 each time, there will be more booklet 1 administered than other booklets.
2. When tests are administered, ensure that attendance lists are filled in. Absent students must be recorded in addition to any irregular issues such as interruptions to test taking. Student participation forms should be used to record attendance of students at testing sessions. Frequently, student participation forms contain student background information (e.g., date of birth, gender) as well as attendance records. The student participating forms must be returned with the test booklets. Absentees of sampled students will have an impact on the sampling weights computation as well as on the overall response rate of participation. Typically, a non-response adjustment is made to the sampling weights when sampled students were not able to take the tests. Figure 4.2 shows an example student participation form.
3. During test administration, instructions to students must be read from a prepared script so that test administration procedures are standardised at different testing locations. This means that test administrators will need to be trained or at least be well informed of testing procedures. It is common to provide a test administrator's manual detailing procedures of conducting the tests. Figure 4.3 shows an example script for the test administrator to read to students.

Overall, test administration needs to be well managed, coordinated and closely monitored. Ensure that the same procedures are followed at all test sites.

School ID: 1045		**School Name: St Mary Primary School**			**Class Name: Grade 3B**	
Student ID	**Student Name**	**Gender**	**Date of Birth**	**Test booklet**	**Attendance**	**comments**
001	Mark Velos	M	30/10/95	6	present	
002	Amy Chen	F	7/11/95	7	absent	
…	…					

Fig. 4.2 An example student participation form

This is a mathematics test. There are ten different test booklets, therefore students around you may be working on different test booklets.
Read each question carefully and answer it as well as you can. Answer as many questions as you can. If you do not know the answer to a question, move on to the next question. You have 60 minutes to work on the test.
Do not start working through the test questions yet. You will be told when to begin.
First we will do some practice questions together. There are five types of questions in the test. …

Fig. 4.3 An example test instruction script to the students

Data Collection and Processing

Capture Raw Data

The first step in gathering data is planning what data to capture. It is essential to capture *raw data* whenever you can. By raw data, we refer to the responses given by the students before any processing is carried out. For example, for a paper-and-pen test, we should capture the options students choose for multiple choice items, rather than capturing the score (correct/incorrect) for each item. We can carry out distractor analysis if we know which options students have chosen, and not just whether students chose the correct option. Further, if there is a mis-key (wrongly specified correct answer), we can easily re-score the data. In most tests we have analysed, there nearly always have been mis-keys.

For short response items, it will be good to capture the actual responses rather than scored responses. For example, for a computation item, it will be good to capture students' answers, and not just correct/incorrect responses. Again, students' incorrect answers can often inform us of misconceptions and help teachers plan remedial lessons.

For extended response items, it may be difficult to apply automatic scoring procedures, so markers are often employed to *categorise* the responses. We deliberately use the word *categorise* rather than *score* here, as we emphasise that the role of the markers is to categorise item responses according to a marking guide, than to decide on scores. Categorising responses according to a well-written marking guide is more objective, while deciding on a score is more subjective as different markers may have different views of giving credits to students. We can always apply scoring rules to the category codes of item responses using a computer program. Make sure a comprehensive marking guide is designed. One should err on the side of providing more coding categories than fewer categories. When marking guides are developed, consider the different responses students may provide and how the responses can be grouped into meaningful categories and codings. For example, in TIMSS, double-digit coding is often used to denote both the score and the response category. Figure 4.4 shows an example TIMSS item with double-digit coding (IEA 2009).

A gardener mixes 4.45 kilograms of rye grass seed with 2.735 kilograms of clover seed to make a mix for sowing a lawn area. How many kilograms of the lawn mix dies he now have? [TIMSS 2007 grade 8 released mathematics item M022046]

Code	Response	Item: M022046
	Correct Response	
10	7.185	
19	Other responses equivalent to 7.185	
	Incorrect Response	
70	6.780 OR 6.78 [4.045 + 2.735]	
71	Contains one miscalculated digit *(e.g., 7.085, 7.195, 8.185 or similar)*	
72	One of the following: 3.18, 31.8, 318, OR 3180 [misaligns decimals]	
79	Other incorrect (including crossed out/erased, stray marks, illegible, or off task)	
	Nonresponse	
99	Blank	

Fig. 4.4 Marking guide for TIMSS released item M022046

It can be seen from Fig. 4.4 that not only the correct answer is captured, but different types of incorrect answers are also captured.

Markers should undergo training to ensure consistency across different markers. Typically, selected student responses to extended-response items, known as control scripts, are distributed to markers during marker training sessions to provide guidelines to markers in categorising student responses according to the marking guide. In Chap. 3, marker harshness/leniency is briefly discussed. Chap. 13 provides some examples of an analysis of marker consistency.

Of course, in designing marking guides, there is always a trade-off between the complexity of coding responses and the amount of information captured. Therefore, a practical approach is to capture as much information as allowed by your available resources. But remember that information not captured will not be available for analysis, while information captured can always be recoded or simplified during the data processing stage. So it is better to err on the side of capturing more information whenever possible. Of course, in practice, one needs to weigh up cost versus benefit. Given the advance of technology now and in the future, extensive data capture and processing will become increasingly cost effective.

Prepare a Codebook

To ensure that there is no ambiguity in capturing data, a codebook should be prepared ahead of data capture. An example codebook for an assessment is shown in Fig. 4.5.

Variable Name	Variable Type	Value	Value Label
Gender	numeric	1	Boy
		2	Girl
		9	Missing
Age	numeric	Between 5-18	Age
		99	Missing
School name	string		
Math Q1	numeric	1	Option A
		2	Option B
		3	Option C
		4	Option D
		9	Missing

Fig. 4.5 An example codebook

For more example codebooks, see TIMSS and PISA websites where codebooks for the databases can be downloaded. A typical codebook provides such information as variable names, variable labels, value coding and the meanings of the codes, as well as valid value ranges. A codebook should provide sufficient information about a data set so that any person analysing the data will know what the data are and how to access them. The release of a data set will be incomplete without a codebook. It is *industry-standard* that each publicly available data set should be accompanied by a codebook.

Data Processing Programs

There are many ways to edit data. A simple way is to edit data in a spreadsheet such as EXCEL. For example, if a student's data need to be deleted, it is easy to highlight a row in EXCEL and press the DELETE key. This process does not require a great deal of programming skills, but there are serious drawbacks to such manual editing of data. First, there is no record of the edits made. Once the data have been changed, there is no record of what changes have been made. Second, if the data set is re-supplied (say, there have been data entry errors, or data scanning errors), which happens frequently, we need to manually repeat the edits all over again. For these reasons, we highly recommend that a computer program is used to carry out all edits of data. In fact, in some organisations specialising in the analysis of assessment data, it has been stipulated that all edits of data must be done using a program script. The program can be any statistical package, for example, SPSS syntax file, SAS program, R program, Microsoft VBA or any other computer language. In the future, there will surely be other programs available. But the key is that all edits can be easily traced, and all edits can be easily repeated if necessary. We also highly recommend that two analysts independently carry out data cleaning and analysis, and compare results. This quality assurance step has proven to be most important, and has frequently been stipulated by professional organisations dealing with data analysis.

Data Cleaning

Data cleaning refers to checking for, and rectifying, anomalies in the data. The types of checks will depend on the specific assessment program, but there are some commonly carried out checks.

- Value range checks. Each data field captured will have an expected range of values. For example, month of birth may take values 1–12; response to a test question may take values 1–5, and so on. The codebook will provide valid value ranges for checking. This is one essential use of the codebook.
- Missing values treatment. It is important to deal with missing responses appropriately. In the context of student testing, there are different types of missing values. First, if there are rotated test booklets so that a student only takes a subset of items, there will be missing item responses because some items are not administered to a student. These missing-by-design responses must not be treated as incorrect answers. Therefore, the code for not-administered items must be separated from other types of missing responses. In carrying out an item analysis, the not-administered items must be treated as missing, and not as incorrect.

  In some assessment programs, a distinction is made between embedded-missing and not-reached items. Embedded-missing refers to items skipped by students, while not-reached items refer to the missing responses at the end of a test, with the possibility that students ran out of time and never had the opportunity to answer the items at the end of a test. Distinguishing between these two types of missing allows the analyst to have the option of applying different treatments to the missing responses. In some cases, not-reached items may be treated as missing and not as incorrect while item difficulty parameters are calibrated. But not-reached items are treated as incorrect when student abilities are computed. In contrast, embedded-missing items are always treated as incorrect. Some rationale for this approach is given as follows. If not–reached items are treated as incorrect, there will be an over-estimation of item difficulty, since the students would answer some correctly had they had the opportunity to answer them. However, for ability estimates, if not-reached items are treated as missing, then students can use a strategy whereby they would answer items sequentially until one item which they do not know the answer. If students stop at this point, and not-reached items are not counted towards the ability estimate, then most students will obtain 100% correct on a test. These are some considerations in deciding on the treatment of missing responses.
- Duplicate record checks. When data are combined from various test administration sites, data could be duplicated inadvertently. On the other hand, data

could be missing as well. So a cross-check should be carried out against the list of sampled schools and students.

- Inconsistency checks. Sometimes, there are survey questions that are inter-linked. For example, in a student questionnaire, there could be a question about the overall number of hours of homework per week. There could also be a question on the number of homework hours for a particular subject, which should not exceed the total number of hours of homework. The relationships between questions on a survey instrument can be used to check for inconsistent answers. Similarly, checks can be made on birth dates if a cohort of students comes from particular grades or age groups. Sometimes, the current year is mistakenly written as the birth year when questionnaires are filled in.
- Multiple instruments checks. In some surveys, a number of instruments are administered. For example, there may be a test on mathematics as well as a student background questionnaire. If these two surveys are placed in separate files, there must be a mechanism for linking the students in both files (e.g., a student ID). Cross-checks should be made between files with linking fields. Care must be taken when files are merged to ensure that merging is carried out correctly, particularly for students with missing instruments.
- If data are manually entered into the computer, a double-entry procedure is often necessary to check for data-entry errors.

Following data checks, errors should be rectified. It is essential to use programming scripts to make changes to data, rather than using interactive data editors to make changes.

Frequently, after item analysis has been run, more data editing needs to take place. For example, a mis-key may be identified from the examination of item discrimination indices, or recoding needs to be carried out for mis-fitting items. Consequently, data preparation and item analysis are often carried out iteratively. It is therefore essential to have an efficient data preparation program in place.

Summary

This chapter lists the key steps in selecting student samples, organising test administration and preparing data for analysis.

For the selection of a sample to represent a population, probability sampling should be used. Typically in large-scale educational assessments, a two-stage sampling procedure is used where schools are first selected and then students are selected from the sampled schools. Two points should be borne in mind. The first is that the sample size needs to be a great deal larger for a two-stage sampling method than for simple random sampling. The second is that the computation and use of sampling weights are very important in all analyses of results.

As test administration often involves multiple testing sites, the organisation of test administration is complex. A clear identification of key personnel involved is important. For example, school principals, teachers and test administrators should all be well-briefed, with supporting documents such as manuals detailing the tasks of each person involved. Test papers need to be printed and delivered, along with student lists and student participation forms.

To prepare for data capture, a codebook of variables should be prepared in advance. Data cleaning and data editing should be carried out programmatically, and not done through interactive data editors. That is, good records need to be kept at every step of the data processing phase. Typically, range checks and missing values coding need special attention. Cross-validation checks should also be carried out. Corrections to data and recoding of variables should all be carried out using program scripts to allow for verification and record keeping.

Discussion Points

(1) Discuss why a two-stage sampling procedure is a great deal less *efficient* than simple random sampling. By *less efficient*, we mean that a larger sample size is needed to achieve the same accuracy.

(2) Discuss why stratification may improve the efficiency of sampling? How should stratification variables be chosen?

(3) In this chapter, a procedure for treating missing values is that not-reached items are treated as missing when item difficulties are estimated. But not-reached items are treated as incorrect when abilities are estimated. Discuss whether this approach will introduce any bias in the results.

(4) With the distribution of rotated test booklets, why shouldn't we always begin with test booklet 1 and assign the booklets sequentially? If there are 30 students per school, and there are 7 rotated test booklets, what is the average number of students taking each test booklet

  (a) if we start with booklet 1 in each school, and
  (b) if we start with a random booklet number between 1 and 7?

(5) Why would absent students have an impact on the computation of sampling weights? How can the sampling weights be adjusted?

Exercises

Q1. The following is an example school questionnaire.

(a) Design a codebook for the data capture for this questionnaire.
(b) List the data cleaning checks you can perform with this questionnaire.

School Questionnaire

Q1 **What type is your school?**

(Put X in one box only)

Private ☐

Government ☐

Q2 **What is the total school enrolment at your school?**

(Write in the number on each row. Please write 0 (zero) if there is none.)

a) ____________ *boys*

b) ____________ *girls*

Q3 **Where is your school located?**

(Put X in one box only)

In a remote area ☐

In a rural area ☐

In or near a small town ☐

In or near a large town or city ☐

Q4 **About how many computers are in the school altogether?**

Number

(Please write 0 (zero) if there is none.) ____________

Q5 **About how many of these computers are available for students to use?** ____________

Q6 **Which of the following does your school have?**

(Put X in one box in each row)

		No	Yes
a)	School assembly hall	☐	☐
b)	First aid room	☐	☐
c)	Music room	☐	☐
d)	Sports field	☐	☐
e)	School canteen	☐	☐

Q7 **How many teachers are there at your school?**

Number

(Please write 0 (zero) if there are none.) ____________

Q8 **How many of these teachers have been at the school for <u>less than 2 years</u>?** ____________

Q9 **How often do the following problems happen with your <u>pupils</u>?**

(Put X in one box in each row)

		Never	Sometimes	Often
a)	Late arrival	☐	☐	☐
b)	Absenteeism	☐	☐	☐
c)	Disturbance and trouble	☐	☐	☐
d)	Health problems	☐	☐	☐
e)	Vandalism	☐	☐	☐

References

IEA (2008) TIMSS 2007 technical report. TIMSS and PIRLS international study center, Lynch School of Education, Boston College, Boston

IEA (2009) TIMSS 2007 user guide for the international database. Released items. Mathematics grade 8. TIMSS and PIRLS international study center, Lynch School of Education, Boston College, Boston

OECD (2009) PISA 2006 technical report. PISA, OECD Publishing, Paris

Further Reading

Cohen AS, Wollack JA (2006) Test administration, security, scoring, and reporting. In: Brennan R (ed) Educational measurement, 4th edn. Praeger publishers, Westport, pp 355–386

Chapter 5
Classical Test Theory

Introduction

Classical Test Theory (CTT), also known as the true score theory, refers to the analysis of test results based on test scores. The statistics produced under CTT include measures of item difficulty, item discrimination, measurement error and test reliability. The term "Classical" is used in contrast to "Modern" test theory which usually refers to item response theory (IRT). The fact that CTT was developed before IRT does not mean that CTT is outdated or replaced by IRT. Both CTT and IRT provide useful statistics to help us analyse test data. Generally, CTT and IRT provide complementary results. For many item analyses, CTT may be sufficient to provide the information we need. There are, however, theoretical differences between CTT and IRT, and many researchers prefer IRT because of enhanced measurement properties under IRT. IRT also provides a framework that facilitates test equating, computer adaptive testing and test score interpretation. While this book devotes a large part to IRT, we stress that CTT is an important part of the methodologies for educational and psychological measurement. In particular, the exposition of the concept of reliability in CTT sets the basis for evaluating measuring instruments. A good understanding of CTT lays the foundations for measurement principles. There are other approaches to measurement such as generalizability theory and structural equation modelling, but these are not the focus of attention in this book.

Concepts of Measurement Error and Reliability

All measurements come with uncertainty (or measurement error), whether the measurements are made in the physical sciences or in the social sciences. There are two types of measurement errors: systematic errors and unsystematic errors. One

M. Wu et al., *Educational Measurement for Applied Researchers*,
DOI 10.1007/978-981-10-3302-5_5

kind of systematic errors is related to validity issues whereby the items in a test consistently measure some latent traits other than those the instrument has been developed to assess. For example, if we use a mathematics test written in English to measure students' mathematics abilities in a non-native English speaking country, then the test scores will not only reflect mathematics abilities but also English proficiencies of non-native English speakers. The test scores from this test will likely have a systematic bias (or error) of underestimating students' mathematics abilities. Another example for possible systematic errors relates to making subjective judgments when scoring students' responses. In this case, the harshness or leniency of a scorer could lead to a systematic bias in estimating the students' abilities.

In contrast, random measurement errors (or unsystematic errors) occur simply due to chance elements. For example, students may make a careless mistake in computation, be distracted momentarily, feel fatigued on a particular testing day, or be lucky in guessing an answer. We also regard the particular set of items selected for a test as due to chance element whereby a student may know more or fewer answers in a test had there been different items in a test. Depending on the type of unsystematic errors, we can assess the degree of variability in test scores due to these errors in different ways. For example, we can administer the same test to the same group of respondents on two different occasions and compute the Pearson correlation coefficient between the two sets of total observed scores. Such a coefficient is known as test-retest reliability. This reliability will capture variations due to carelessness, fatigue on a day and other random errors due to the testing occasion, but this reliability will not capture the variation in test scores due to the selection of items in a test. Consequently, there are different types of reliability measures targeting different sources of random errors.

In this book, we are particularly interested in measurement errors due to the selection of test items, as we believe this source contributes the most variation in test scores. We can think of a test as an instrument used to sample students' knowledge/skills in a subject domain. Because a test has limited number of items, we typically only obtain a rather small sample of each student's capabilities through administering, say, a one-hour test. Should students sit another test with different items but testing the same construct, the test scores of students are likely to vary across the two tests. We provide an example to illustrate this below.

Suppose we develop tests to measure grade 5 students' mathematics abilities. Imagine there is a large item pool of all possible grade 5 mathematics items. We take 40 items from this item pool to make up a test (see Fig. 5.1).

David is a grade 5 student. If we have the opportunity to administer all possible grade 5 mathematics items to David, David can answer 60% of the items correctly. Of course in real-life we will not likely have the opportunity to do this owing to the time and effort required. When a test is made up of 40 items, David's expected score is 24 (60% of 40). But there is a chance element of the items selected in a test, so David's actual test score may not be exactly 24. Suppose three tests are constructed (labelled 2008, 2009 and 2010 in Fig. 5.1) by selecting 40 questions from the item pool. David takes these three tests one day and obtains 20, 28 and 25.

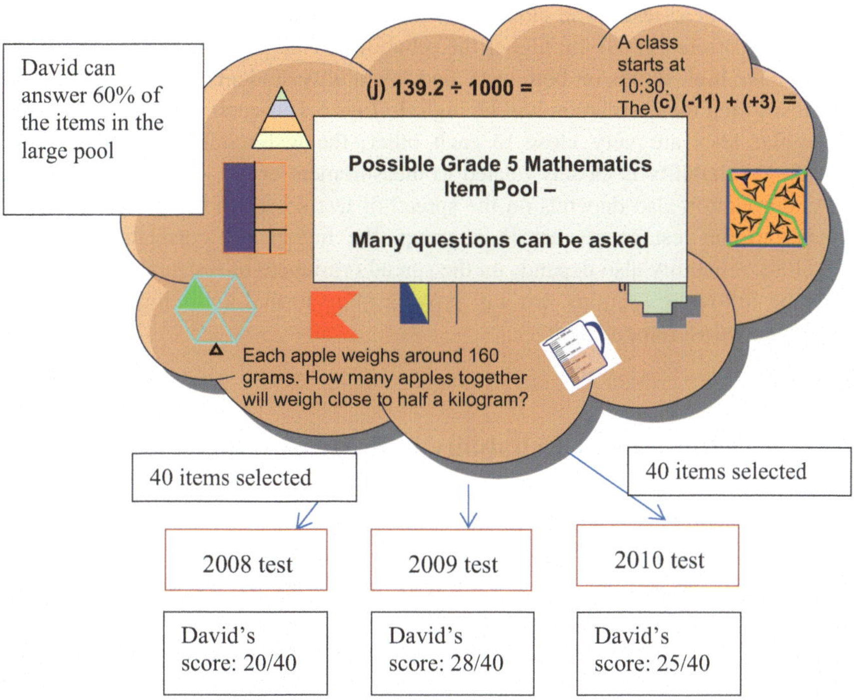

Fig. 5.1 Test scores on parallel tests

That is, there is a variation in David's test scores on similarly constructed tests known as parallel tests (see a formal definition of parallel tests in the latter part of this chapter). From our past experience in analysing large-scale assessments, we estimate that David's test scores will likely vary between 20 and 30 out of a possible maximum of 40, given that David knows 60% of the grade 5 mathematics content.

This variation in David's test scores on similar tests (or parallel tests) relates to the concept of measurement error. That is, there is a margin of error in measuring David's mathematics ability using a (somewhat short) test. It is unfortunate that the word "error" is used, since "error" suggests mistakes. Measurement error is not pertaining to mistakes in setting the test questions or mistakes in scoring the answers, but measurement error comes from the fact that we have only collected a small sample of a student's capabilities when a test is administered. Of course if there are mistakes in setting the questions and in processing the results, the measurement error will increase. But predominantly, measurement error comes from the fact that limited number of items in a test is administered.

Suppose there is a group of students who all took the 2008 and 2009 tests (at the same point in time so student abilities haven't changed in between taking the two tests). So each student has a pair of test scores. If we compute the correlation between the pair of test scores across all students, this correlation reflects the reliability of these tests. The closer the values of each pair of test scores are, the higher the correlation, so the higher the reliability. That is, if high ability students tend to have high scores on both tests, and low ability students have low scores on both tests, then the correlation will be high. In Fig. 5.1, if each student's test scores on similar tests are very close to each other, the test reliability will be high. Therefore, reliability is closely related to measurement error.

As correlation also depends on the spread of test scores in the group, the more spread out the test scores are in a group, the higher the correlation will be. Therefore, reliability also depends on the spread (variance) of test scores in a group.

In the following sections, we will provide more formal definitions of test reliability and measurement error.

Formal Definitions of Reliability and Measurement Error

The term reliability was first coined by Charles Spearman in 1904. It has been an area of active research ever since by measurement researchers in many fields. This section will give a brief account of reliability from the classical test theory approach.

Classical test theory makes the assumption that each respondent's observed score on a test is the sum of his/her "true score" and an error score, as expressed by Eq. (5.1):

$$X_n = T_n + E_n \tag{5.1}$$

where X_n, T_n and E_n stand for the observed, true and error scores, respectively, The subscript n refers to the nth respondent. The true score, T_n, is defined as the average of test scores if a test is repeatedly administered to a student (and the student can be made to forget the content of the test in-between repeated administrations). Mathematically, the definition for the true score is $E(X_n)$, the expectation of the observed scores, where the expectation is taken over repeated administrations of the same test to the same student. The true score is assumed to be a stable measure reflecting a student's level on the construct being measured, while the error score is assumed to be an unsystematic error or random error.

Assumptions of Classical Test Theory

First, we define $X = (X_1, X_2, \ldots, X_N), T = (T_1, T_2, \ldots, T_N)$ and $E = (E_1, E_2, \ldots, E_N)$ as observed, true and error scores, respectively, across all N students. Five classical test theory assumptions are listed below.

(1) The observed score for a student is the sum of the true score and error score $X_n = T_n + E_n$.
(2) Since by definition, $T_n = E(X_n)$, it follows that $E(E_n) = 0$. That is, the expectation of the error scores over repeated administrations of a test for each student is zero.
(3) The correlation between the error score and true score is 0. That is, $corr(T, E) = 0$. This means that students with high true scores will not consistently have higher or lower error scores.
(4) Let Test A and Test B be two tests administered to the same students. The true scores on Test A (T_A) is not correlated with the error scores on Test B (E_B). That is, $corr(T_A, E_B) = 0$.
(5) Let Test A and Test B be two tests administered to the same students. The error scores on Test A (E_A) is not correlated with the error scores on Test B (E_B). That is, $corr(E_A, E_B) = 0$.

Definition of Parallel Tests

Test A and Test B are said to be parallel tests if they have observed scores X and X' that satisfy Assumptions 1–5 above, and $T = T'$, $Var(E) = Var(E')$. Note that in the introductory sections on reliability in this chapter, we used the term "similar tests" to illustrate the idea of reliability. These similar tests are meant to be "parallel" as parallel tests are defined here.

Definition of Reliability Coefficient

Test reliability can be defined in a number of ways. We will use one definition here to begin with, and then show that other definitions will follow from this definition when the above five assumptions are satisfied. We define the reliability of a test as the correlation between the observed scores on this test and observed scores on a parallel test. Mathematically, we write

$$\rho_{XX'} = corr(X, X') \tag{5.2}$$

as the definition for test reliability. Using Assumptions 1–5 above and the definition of parallel tests, it can be shown that

$$\begin{aligned}
\rho_{XX'} &= corr(X, X') \\
&= \frac{\operatorname{cov}(X, X')}{\sqrt{Var(X)Var(X')}} \\
&= \frac{\operatorname{cov}(T+E, T'+E')}{\sqrt{Var(X)Var(X')}} \\
&= \frac{\operatorname{cov}(T, T') + \operatorname{cov}(T, E') + \operatorname{cov}(E, T') + \operatorname{cov}(E, E')}{\sqrt{Var(X)Var(X')}} \\
&= \frac{\operatorname{cov}(T, T')}{\sqrt{Var(X)Var(X')}} \\
&= \frac{\operatorname{cov}(T, T)}{\sqrt{Var(X)Var(X)}} \\
&= \frac{Var(T)}{Var(X)}
\end{aligned} \tag{5.3}$$

That is, another interpretation of test reliability is that it is the proportion of the variance of the true scores out of the variance of the observed scores.

Further, note that since $X_n = T_n + E_n$,

$$\begin{aligned}
Var(X) &= Var(T) + Var(E) + 2\operatorname{cov}(T, E) \\
&= Var(T) + Var(E)
\end{aligned} \tag{5.4}$$

Therefore the test reliability can also be expressed as

$$\rho_{XX'} = \frac{Var(T)}{Var(T) + Var(E)} \tag{5.5}$$

When test reliability is high, one would expect that the correlation between the observed scores and true scores to be high as well. An examination of the correlation between the observed and true scores shows the following:

$$\begin{aligned}
corr(X, T) &= \frac{\operatorname{cov}(X, T)}{\sqrt{Var(X)Var(T)}} \\
&= \frac{\operatorname{cov}(T+E, T)}{\sqrt{Var(X)Var(T)}} \\
&= \frac{\operatorname{cov}(T, T) + \operatorname{cov}(E, T)}{\sqrt{Var(X)Var(T)}} \\
&= \frac{Var(T)}{\sqrt{Var(X)Var(T)}} \\
&= \sqrt{\frac{Var(T)}{Var(X)}} \\
&= \sqrt{\rho_{XX'}}
\end{aligned} \tag{5.6}$$

Consequently,

$$\rho_{XX'} = (corr(X,T))^2 \tag{5.7}$$

That is, the reliability is equal to the square of the correlation between observed scores and true scores.

So we have four alternative ways of interpreting test reliability as given by Eqs. (5.2), (5.3), (5.5) and (5.7). For further details on the assumptions and derivations of classical test theory and reliability, see Allen and Yen (1979) as well as Brennan (2006).

Computation of Reliability Coefficient

In practice, we do not know the true scores, *T*. Therefore, the definitions of reliability as given by Eqs. (5.3), (5.5) and (5.7) are not very useful to us for actually computing the reliability. Equation (5.2) can possibly be used if we construct parallel tests. Nevertheless it will be quite time-consuming to construct and administer multiple tests in order to compute the test reliability.

If we have only one set of observed scores, an intuitive treatment is to somehow split the set of items into two halves, for example using an odd-even item number split, and then take the Pearson correlation coefficient between the two halves of the test. This treatment can be regarded as an extension of the idea of parallel test forms. The resulting coefficient is known as the split-halves coefficient and it reflects how internally consistent the items of the test are. Since the number of items of a single test is being divided into two halves of smaller test, the split-halves coefficient may underestimate the reliability of the test of the original length. A usual practice is to apply the Spearman-Brown prophecy formula (Spearman 1910; Brown 1910) to project what the coefficient would be if items similar to those in the original test were added so that the number of items in each half amounts to the same as in the original test (see Additional Notes).

Since there are many possible ways to split the number of items into two halves other than the odd-even split, a reasonable idea is to split the test in all possible way and then take the average of all the reliabilities of each possible split. Though not being practical if the split-halves are physically carried out, it can be shown that the Cronbach's alpha coefficient (Cronbach 1951) serves as a good estimate to the mean reliability of all possible splits. Cronbach's alpha coefficient is expressed as

$$\alpha = \frac{I}{I-1}\left(\frac{Var(X) - \sum_{i=1}^{I} Var(item\ score\ of\ item\ i)}{Var(X)}\right) \tag{5.8}$$

where $Var(X)$ refers to the variance of the test scores across students, and I is the total number of test items. An intuitive way to understand coefficient α is to think of

$\sum_{i=1}^{I} Var(item\ score\ of\ item\ i)$ as $Var(E)$ so that $Var(X) - Var(E) = Var(T)$ in the numerator of the expression in brackets of Eq. (5.8). Then Eq. (5.8) becomes very similar to Eq. (5.3) for the definition of reliability. It can be shown that Eq. (5.8) provides a lower bound for the test reliability, $\rho_{XX'}$. That is, $\rho_{XX'} \geq \alpha$.

Interested readers can refer to Cronbach (1951), Nunnally and Bernstein (1994) for details. Another approach to estimate the mean reliability of all possible splits is the KR-20 coefficient reported in Kuder and Richardson (1937). It turns out that the KR-20 coefficient can be regarded as a special case of the Cronbach's alpha coefficient when the latter is applied to dichotomously scored data. Specifically, for dichotomous items,

$$\alpha = \frac{I}{I-1}\left(\frac{Var(X) - \sum_{i=1}^{I} p_i(1-p_i)}{Var(X)}\right) \tag{5.9}$$

where p_i is the proportion of students obtaining the correct answer on item i.

The split-halves, the KR-20 and the Cronbach's alpha coefficients all yield information about the internal consistency of a test based on a single test administration.

There is much work done in the area of reliability. For further reading, we recommend two books. The first one is by Traub (1994), which gives an accessible introduction account on reliability. The second one is written by Knapp (2009), which is written under a conversational tone and gives a fairly comprehensible account on many issues related to reliability.

Additional Notes on Spearman-Brown Prophecy Formula

One important factor influencing measurement error is the test length (i.e., the number of items in a test). If the test length of a new test is k times the test length of the original test, then the variance of the error scores for each student will be *1/k* of the error variance of the original test. That is, $Var(E') = \frac{Var(E)}{k}$. It can be shown that the test reliability, $\rho_{YY'}$, of the new test can be expressed as the reliability of the original test, $\rho_{XX'}$, as

$$\rho_{YY'} = \frac{k\rho_{XX'}}{1 + (k-1)\rho_{XX'}}$$

If a new test is twice the length of an original test, then the reliability of the new test can be predicted from the reliability of the original test as

$$\rho_{YY'} = \frac{2\rho_{XX'}}{1 + \rho_{XX'}}$$

The above formula is generally known as the Spearman-Brown split-half reliability coefficient (Spearman 1910; Brown 1910).

Standard Error of Measurement (SEM)

From Eqs. (5.3) and (5.4), we can derive an expression for $Var(E)$ in terms of reliability, $\rho_{XX'}$:

Since $\rho_{XX'} = \frac{Var(T)}{Var(X)} = \frac{Var(X) - Var(E)}{Var(X)}$, by re-arranging the terms, we obtain

$$Var(E) = (1 - \rho_{XX'})Var(X)$$

Taking square-roots on both sides, we have

$$\begin{aligned} \text{standard error of measurement} &= \sqrt{Var(E)} \\ &= \sqrt{(1 - \rho_{XX'})Var(X)} \end{aligned}$$

Standard error of measurement can be used to provide a degree of uncertainty in test scores. For example, if a test has a reliability of 0.8 and the variance of the test scores across students is 45, then the standard error of measurement is $\sqrt{(1 - 0.8) \times 45} = 3$. If a student's test score is 20, we can make an inference that 95% of the student's test scores on parallel tests are likely to be in the range of $(20 - 2 \times 3,\ 20 + 2 \times 3)$, that is, in the range of (14, 26). Note that the standard error of measurement computed in this way applies to all observed scores in a test.

Correction for Attenuation (Dis-attenuation) of Population Variance

Given that $\rho_{XX'} = \frac{Var(T)}{Var(X)}$, by re-arranging the terms, we obtain

$$Var(T) = \rho_{XX'} \times Var(X).$$

So after we have obtained the reliability and computed the variance of the observed scores, we can estimate the variance of the true scores by multiplying the reliability to the variance of the observed scores. In this way, we can estimate the variance of the true scores even though we do not know each student's true score. This process is commonly known as correction for attenuation (or dis-attenuation) of measurement error for the population variance. The variance of the observed scores is always larger than the variance of the true scores, because the measures are "attenuated" by measurement error.

Correction for Attenuation (Dis-attenuation) of Correlation

When each student takes two tests, the correlation between the two test scores across students will be lower than the correlation of students' true scores, because each test score contains measurement error. This observed correlation is said to be attenuated by measurement error. A correction for the observed correlation can be made using the test reliabilities of the two tests, as shown below.

$$corr(T_1, T_2) = \frac{corr(X_1, X_2)}{\sqrt{R_1 R_2}},$$

where X_1, X_2 are the observed test scores on Test 1 and Test 2 respectively; T_1, T_2 are the true scores on the two tests, and R_1, R_2 are the test reliabilities for Tests 1 and 2 respectively. It can be seen that if the test reliabilities are low, then the correction factor $1/\sqrt{R_1 R_2}$ will be large. On the other hand, if the test reliabilities are high (i.e., close to 1), then the correlation of observed test scores will be close to the correlation of true scores.

Other CTT Statistics

In this section, we introduce a number of CTT statistics typically used in item analysis, including item difficulty and item discrimination measures. To illustrate these statistics, we use an example data set containing dichotomously scored item responses to 10 mathematics items from 400 students. The items are shown in Table 5.1.

Table 5.2 shows the first 9 lines of scored item responses in the data file.

Item Difficulty Measures

Under CTT, item difficulty measures are simply the percentages of students obtaining the correct answer. In our example, we compute the percentage of a score of "1" for every column in Table 5.2. The results are given in Table 5.3.

Table 5.3 shows that 63% of the students obtained the correct answer for Q1; 76% for Q2, etc. It can be seen that Q5 is the easiest question on this test, being a straightforward computation item. In contrast, Q10 is the most difficult item, where the wording (one-fifth) and the answer format may be unfamiliar to some students.

Under CTT, the item difficulty measure is simply the proportion correct for an item. This is a very intuitive measure of the item difficulty. When an item has partial credit scoring, there is a slight modification to the computation of proportion or percentage correct. Suppose an item has possible scores of 0, 1 and 2. The proportion correct for each score category is shown in Table 5.4.

Table 5.1 Ten mathematics test items

1	If 2nd of May is a Friday, what day of the week is 20th of May?
2	The scale of a map is 1 cm for 25 km. If the distance on the map between point A and point B is 3 cm, what is the real distance?
3	The floor of one room is 5 m wide and 7 m long. What is the total floor area?
4	What is the place value of 6 in 26, 344?
5	652 – 184 = ?
6	76 ÷ 4 = ?
7	A farmer has 63 chickens. 7/9 of them are hens and the remaining are roosters. How many hens are there?
8	Find the value represented by the symbol "?" in the following equation: 24–8 = 4 × (1 + ?)
9	Which set of numbers is in order from the smallest to the largest: A. (257, 311, 401); B. (422, 337, 498); C. (265, 322, 299); D. (396, 383, 400)
10	One-fifth of 36 is between: A. 4 and 5; B. 5 and 6; C. 6 and 7; D. 7 and 8

Table 5.2 Excerpt of the item response data for the ten math items

	Q1	Q2	Q3	Q4	Q5	Q6	Q7	Q8	Q9	Q10	Test score
Student 1	1	0	0	1	1	0	0	0	1	0	4
Student 2	0	0	0	0	1	0	0	0	1	0	2
Student 3	0	0	0	0	1	0	0	0	1	0	2
Student 4	0	0	1	0	0	0	0	0	0	0	1
Student 5	0	1	1	1	1	1	1	1	0	0	7
Student 6	1	1	0	1	0	0	1	0	1	0	5
Student 7	0	1	0	0	1	1	0	0	0	0	3
Student 8	1	1	0	0	1	1	1	0	0	0	5
Student 9	0	1	0	1	1	0	0	0	1	0	4
	…	…	…	…	…	…	…	…	…	…	…

Table 5.3 Proportion correct for the ten items in the example

Q1	Q2	Q3	Q4	Q5	Q6	Q7	Q8	Q9	Q10
0.63	0.76	0.79	0.79	0.91	0.88	0.65	0.63	0.81	0.37

Table 5.4 Proportion correct for item categories of a hypothetical partial credit item

Score 0	Score 1	Score 2
0.23	0.60	0.17

Table 5.4 shows that 23% of the students obtained a score of 0 on this item; 60% obtained 1 and 17% obtained 2. Therefore, the average score on this item is computed as $0 \times 0.23 + 1 \times 0.60 + 2 \times 0.17 = 0.94$. That is, the observed average score on this item is 0.94 out of a possible maximum of 2. Therefore, expressed as a "proportion" ranging between 0 and 1, we divide 0.94 by 2. The "percentage correct" for this item is then $0.94 / 2 = 0.47$. This makes sense as we look through Table 5.4 and find that more than half of the students obtained the middle score; about one-fifth obtained 0 or 2, so a "percentage correct" for the item should be around the 50% mark. More generally, to compute the "percentage correct" for a partial credit item, we can use the formula

$$\text{percentage correct} = \frac{\sum_{k=0}^{K} k n_k}{K \sum_{k=0}^{K} n_k} \tag{5.10}$$

where the capital letter, K, stands for the maximum score for an item; n_k is the number of students in each score category, k. The numerator in Eq. (5.10) is the observed score on this item summed over all students, while the denominator is the maximum possible score assuming all students answer the item correctly.

Item Discrimination Measures

Item discrimination is a measure of the relationship between the score on an item and the overall test score. In the last column of Table 5.2, a test score is computed for every student. This test score can be regarded as a measure of the underlying construct; in this case it is the mathematics ability. If students' scores on an item are closely related to their mathematics abilities (as measured by the test scores), then a correlation between the item score and the test score will be high. That is, for students who have high test scores, their item scores on this item are more likely to be 1. For students with low test scores, their item scores are more likely to be 0. For example, if we compute the correlation between column 1 (Q1) and column 11 (test score) of Table 5.2, we obtain 0.54, indicating that there is a good positive linear relationship between Q1 score and the total score.

Strictly speaking, when the correlation between item score and test score is computed, the test score should not include the item under consideration. Since the test score in column 11 of Table 5.2 includes Q1 score, a correlation between the two scores will be inflated. Therefore, for Q1, we first compute the test score excluding Q1. Then, we compute the correlation between this revised test score and Q1 score. We obtain a correlation coefficient of 0.39. Table 5.5 shows the correlation between item score and test score (excluding item under consideration) for all 10 items.

The correlation coefficients shown in Table 5.5 are also known as the point-biserial correlation (pbis, or pb). In this book, we will use the term "item

Table 5.5 Correlation coefficient between item score and test score (excluding item) for the ten items in example

Q1	Q2	Q3	Q4	Q5	Q6	Q7	Q8	Q9	Q10
0.39	0.59	0.52	0.45	0.40	0.48	0.64	0.60	0.51	0.36

discrimination index" to refer to the relationship between item score and test score. It is a measure of how well an item score is associated with test score. Imagine we ask a question unrelated to mathematics ability in a test, such as "Is blue your favourite colour?", and score 1 for yes, 0 for no. If we compute the correlation between the score on this colour preference question and test score in Table 5.2, we would expect a correlation close to zero. That is, this question is not related to measures of mathematics ability. Consequently, we can use the magnitude of the item discrimination index to assess how well an item is tapping into the latent construct.

There are at least two possible reasons for poorly discriminating items. The first is that an item tests something else compared to the majority of items in the test. The second is that an item is poorly written and confuses students. Whenever we examine low discrimination items, we should first check whether the wording and format of the item is problematic, and then check whether the item may be testing a different construct than that intended for the test.

In our example, we find that there is some variation in item discrimination across the 10 items. Q1 and Q10 have the lowest discrimination indices while Q7, Q8 and Q2 have relatively higher discrimination indices. On examining the items, some conjectures and observations can be made. The wording and answer format of Q10 may confuse students of all ability levels. Q1 appears to be a straightforward counting problem. But such counting can easily lead to careless mistakes so that occasionally higher ability students may obtain the wrong answer through a slip in counting, thus lowering the discrimination index. In contrast, Q7, Q8 and Q2 are textbook style word problems. They are not too easy or too difficult (percentages correct are 65, 63 and 76% respectively), but by and large higher ability students are more likely to obtain the correct answers for these three questions.

Item Discrimination for Partial Credit Items

When an item has partial credit scoring such as 0, 1 and 2, we can still compute the correlation between item score and test score. In this case, because the item score has more score points, the correlation will generally be higher. That is, the item score can divide students into three groups (0, 1 and 2) instead of two groups (0 and 1), potentially providing more power in discriminating students.

When there are more than two item score categories, we can also form a correlation for individual score categories. For example, column 2 of Table 5.6 shows an example of scored partial credit responses.

Table 5.6 Illustration of point-biserial correlation coefficients for item score category 0 in a partial credit item

Student	Item score	Indicator variable for score category 0	Test score
1	0	1	18
2	1	0	20
3	1	0	24
4	2	0	30
5	0	1	15
6	2	0	25
7	1	0	22

For each score category (0, 1 and 2), we can compute a point-biserial correlation index by first creating an indicator variable to indicate whether the response is in the score category or not. Table 5.6 shows an example for creating an indicator variable for score category 0. We create an indicator variable (third column of Table 5.6) corresponding to the item scores in column 2. Whenever the item score is 0, we assign 1 to the indicator variable. For all other score categories (1 and 2 in this case), we assign 0 to the indicator variable (compare columns 2 and 3 in Table 5.6). Correlation is then computed between the indicator variable and the test score (columns 3 and 4 in Table 5.6). Since score category 0 is now recoded to 1 in the indicator variable, and score categories 1 and 2 are recoded to 0, we expect the correlation between the indicator variable and the test score to be negative. That is, when the overall test score is low, we expect the indicator variable to have a value of 1. When the overall test score is high, we expect the indicator variable to have a value of 0.

Similarly we can carry out recodes for score category 1 by recoding item scores of 0 and 2 to 0, and keeping category 1 as 1, then compute the correlation between the recoded score and test score. For score category 2, we recode the item score of 2 to 1, and item scores of 0 and 1 to 0. In summary, when we compute the point-biserial correlation for a score category, we recode that category to 1, and all other categories to 0. Table 5.7 shows an example output of discrimination indices at the item level, and point-biserial correlations at item response category level.

Table 5.7 Example item discrimination and category point-biserial for a partial credit item

Item discrimination = 0.70	
Item response category	Point biserial
0	−0.65
1	−0.06
2	0.60

Distinguishing Between Item Difficulty and Item Discrimination

Conceptually, the notion of item difficulty and item discrimination should be clearly separated. That is, whether an item is difficult or easy, it should be independent of whether an item is discriminating or not. Although in practice, as CTT item discrimination is computed as a correlation coefficient between item score and test score, this correlation is related to the item difficulty. For example, if an item is very easy, then the majority of item score is 1. Consequently, any correlation using the item score will not likely produce high correlation. Similarly for very difficult items, the correlation is likely low if most students' item scores are 0. Therefore, when low discriminating is observed, one may check whether the item is very easy or very difficult, which may be a cause for low discrimination. Nevertheless, we would like to stress interpreting item difficulty and item discrimination as different concepts.

When an item is very easy or very difficult, the item may still be useful to be included in a test. Depending on the purposes of a test, we may want to include some very easy and very difficult items. For example, frequently, we place one or two easy items at the beginning of a test to ease any anxiety on the part of the students. Further, if a test is an all-purpose test targeting a population with a wide range of abilities, then there needs to be a wide range of item difficulties to measure low and high ability students. Therefore, a very difficult item may still have a place in a test, provided that the handful of students who answered the item correctly are of high abilities. So that brings us to the notion of discrimination.

It is much more problematic when an item does not discriminate between low and high ability students, irrespective of the item difficulty. Such an item will lower the test reliability, increase measurement error, and make the test scores less interpretable. From the point of view of measurement ideals (discussed in more detail in Chap. 6), it is highly desirable to include items that discriminate well between low and high ability students. If we include some poorly discriminating item in a test because "the items test important concepts", then we are departing from the principles of the ideal measurement discussed in this book. In short, you can have a test consisting of items not highly correlated with each other, but the test scores will not be easily interpretable, and you will not have desirable measurement properties valued by many measurement proponents. In such cases, the test construct should be re-examined to consider the division of the test into several tests, each testing a central construct. This is a recommendation from a theoretical viewpoint. Of course in practice we are faced with many obstacles in achieving good measurement. For example, in education a test is typically based on a curriculum. Curricula are not typically designed around measurement principles and latent constructs. Therefore, we are often faced with the tension between content validity (what to include in a test to cover the curriculum) and good properties of measurement (whether test scores reflect a single construct).

To sum up about item difficulty and item discrimination, we will state that item difficulty is about how many students obtained the correct answer, while item discrimination is about which students obtained the correct answer (high or low ability students). Keeping these two concepts in mind will assist us with making sense of item statistics.

Discussion Points

(1) Discuss the factors influencing test reliability. Consider factors including the sample size of respondents, the test length, the quality of the items, the variation in respondents' abilities and test targeting.
(2) What is the impact of measurement error on test reliability? Does measurement error provide the same information as test reliability?
(3) What factors have an impact on the point-biserial correlation coefficients of test items? How do items with low point-biserial correlations impact on test reliability?

Exercises

Q1. The following demonstrates a simulated dataset of 20 students' true scores and their raw scores on a 10-item test. For example, person #1 with a true score of 0.7 indicates that student #1 can response correctly 70% of the items in a large item pool. If responses to the 10 items in the test are randomly drawn from the item pool, then the observed scores for each item for student #1 can be seen as a random draw from a binomial distribution with a probability of success equal to 0.7. Carry out such a simulation to generate observed scores for 200 students. Compute the correlation between true scores and observed scores. Also compute Cronbach's alpha coefficient. Compare the results. Generate another set of observed scores using the same true scores. Compute the correlation between the observed scores from both data sets. How does this correlation coefficient compare with Cronbach's alpha coefficient?

The raw scores of a 10-item test for 20 students

Student ID	True score	Item number										Observed score
		1	2	3	4	5	6	7	8	9	10	
1	0.7	1	1	1	1	1	0	1	1	0	0	7
2	0.8	0	1	0	1	1	1	1	0	1	1	7
3	0.4	0	1	0	1	0	0	0	0	1	1	4
4	0.7	1	1	0	1	1	1	1	1	1	1	9
5	0.9	1	0	1	1	1	0	0	1	1	0	6
6	0.5	0	1	1	1	0	0	0	1	1	0	5

(continued)

(continued)

Student ID	True score	Item number										Observed score
		1	2	3	4	5	6	7	8	9	10	
7	0.3	0	0	0	1	1	0	0	0	1	1	4
8	0.3	0	0	1	0	0	1	0	0	0	0	2
9	0.7	1	0	1	1	1	0	1	0	1	1	7
10	0.8	0	1	1	1	1	1	1	1	0	1	8
11	0.3	1	1	0	0	0	1	1	0	0	0	4
12	0.2	0	0	0	0	0	0	0	0	0	0	0
13	0.2	0	1	1	0	0	0	0	0	0	0	2
14	0.5	1	0	0	1	0	0	0	1	1	0	4
15	0.9	1	1	1	1	1	1	1	1	1	1	10
16	0.2	0	0	1	0	0	0	0	1	0	0	2
17	0.8	1	0	1	1	1	1	0	1	1	1	8
18	0.5	0	1	1	1	1	0	0	1	1	0	6
19	0.1	1	0	0	0	0	1	0	0	0	0	2
20	0.4	0	0	0	1	0	0	1	0	1	0	3

Q2. Use simulation to generate item response data of various test length keeping the sample size of respondents constant. Compute test reliability as a function of test length. Plot the graph.

Q3. Use simulation to compute test reliability as a function of sample size of respondents, keeping the test length constant. Plot the graph.

Q4. Use simulation to compute test reliability for a class of high achievers and for a class of mixed ability students. Compare the reliabilities for the test taken by these two classes.

References

Allen MJ, Yen WM (1979) Introduction to measurement theory. Brooks/Cole Publishing Company, Monterey

Brennan RL (ed) (2006) Educational measurement, 4th edn. Praeger Publishers, Westport

Brown W (1910) Some experimental results in the correlation of mental abilities. Br J Psychol 3:296–322

Cronbach LJ (1951) Coefficient alpha and the internal structure of tests. Psychometrika 16: 297–334

Gulliksen H (1950/1987) Theory of mental tests. Lawrence Erlbaum Associates, Mahwah, NJ

Knapp T (2009) The reliability of measuring instruments. Available at http://tomswebpage.net/

Kuder G, Richardson M (1937) The theory of estimation of test reliability. Psychometrika 2: 151–160

Lord FM, Novick MR (1968) Statistical theories of mental test scores. Addison-Wesley, Reading

Nunnally JC, Bernstein IH (1994) Psychom theory, 3rd edn. McGraw-Hill, New York

Spearman C (1904) The proof and measurement of the association between two things. Am J Psychol 18:160–169
Spearman C (1910) Correlation calculated from faulty data. Br J Psychol 3:271–295
Traub R (1994) Reliability for the social sciences: theory and applications. Sage, Thousand Oaks

Further Reading

Crocker L, Algina J (1986) Introduction to classical and modern test theory. Wadsworth, Belmont
Ghiselli EE, Campbell JP, Zedeck S (1981) Measurement theory for the behavioural sciences. Freeman and Company, San Francisco

Chapter 6
An Ideal Measurement

Introduction

When one undertakes the measurement of a latent trait, what are the desirable properties one would like to have for the measures? Clearly, reliability and validity are important considerations, as discussed in the preceding chapters. If the measures are not reliable, we cannot put a great deal of trust in the scores of measurement. If the measures are not valid, we cannot provide a great deal of interpretation for our measures. Apart from reliability and validity, what are other properties of good measurements? Chapter 1 presents a discussion about levels of measurement, and suggests that if interval or ratio levels of measurement can be achieved, the measures will be better than ordinal measurement. In this chapter, we take a closer look at good properties of measurement, and present an approach that will attempt to achieve several desirable properties of measurement.

An Ideal Measurement

Consider an example where one is interested in measuring students' academic ability in a subject domain. Suppose a test is developed for this purpose, one would like the test scores to be accurate and useful.

By accurate, we mean that the score a student obtains can be "trusted". That is, if Tom gets 12 out of 20 on a geometry test, we hope that this score provides a measure of what Tom can do on this test, and that if similar tests could be administered, he is likely to get 12 out of 20 again. This notion of "accuracy" relates to the concept of "reliability" in educational measurement.

We would also like the test scores to be useful for some purpose we have in mind. For example, if we want to select students for a specialist course, we would

M. Wu et al., *Educational Measurement for Applied Researchers*,
DOI 10.1007/978-981-10-3302-5_6

want our test scores to reflect students' suitability for taking this course. This notion of "usefulness" relates to the concept of "validity" in educational measurement.

Furthermore, we would like the test scores to provide us with a stable frame of reference in comparing different students. For example, if the test scores from one test tell us that, on a scale of geometry ability from low to high, Tom, Bev and Ed are located as follows:

If we give Tom, Bev and Ed another test on geometry, we hope that they will be placed on the geometry ability scale in the same positions relative to each other as that shown in Fig. 6.1. That is, no matter which geometry test is administered, the result will show that Bev is a little better than Tom in geometry, but Ed is very much better than both Tom and Bev. If this can be achieved, the measurement is at the interval level, where statements about the distances between students can be made, and not just rank ordering. The measurement also has an "invariance" property in that the placement of students on the ability line does not change when different tests tapping into the same construct are administered. In the following section, we will identify some problems with using test scores as ability estimates, in relation to the measurement invariance property.

Ability Estimates Based on Raw Scores

Let us consider using raw scores on a test as a measure of ability. The term "raw" refers to that the test scores have not been transformed in any way. Suppose two geometry tests are administered to a group of students, where test 1 is easy and test 2 is hard. Suppose A, B, C and D are four students with differing abilities in geometry. A is an extremely able student in geometry, B is an extremely poor student in geometry, and C and D are somewhat average students in geometry.

If the scores of students A, B, C and D on the two tests are plotted, one may get the four points shown in Fig. 6.2.

From Fig. 6.2, one can see that student A, being excellent in geometry, is likely to score high on both the easy test and the hard test. Student B, being rather poor at geometry, is likely to score low on both tests. Students C and D are likely to score somewhat higher on the easy test, and somewhat lower on the hard test.

On the horizontal axis where the scores on the easy test are placed, it can be seen that A and C are closer together than B and C in terms of their raw scores. However, on the vertical axis where the scores on the hard test are placed, A and C are further

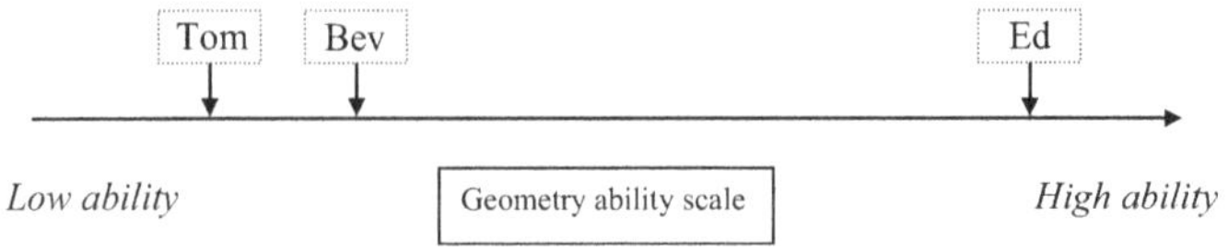

Fig. 6.1 Locations of Tom, Bev and Ed on the geometry ability scale

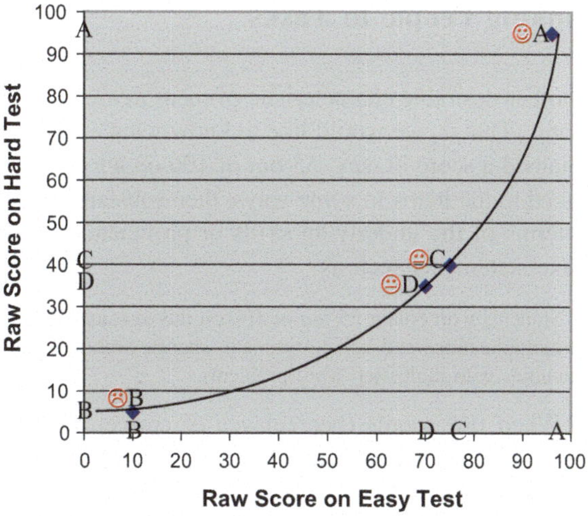

Fig. 6.2 Plot of student raw scores on an easy test and a hard test

apart than B and C. If both the easy test and the hard test measure the same ability, one would hope to see the same distance between A and C, irrespective of which test is administered. From this point of view, we can see that raw scores do not provide us with a stable frame of reference in terms of the distances between students on the ability scale. However, raw scores do provide us with a stable frame of reference in terms of ordering students on the ability scale.

In more technical terms, one may say that, in this example, raw scores provide ordinal measurement, and not interval measurement. Consequently, at least in some cases, the ability estimates based on raw scores are dependent on the particular test administered. This would not be a desirable characteristic of an ideal measurement.

However, the example we provided is quite an extreme case. In practice, if the test difficulties are similar across different tests and the tests are long, raw scores can provide near-interval measures, particularly for the ability range in the centre of the distribution. It may be better to say that raw scores provide measures somewhere in-between ordinal and interval measurement. For example, from Fig. 6.2, one can still make the judgement that C and D are closer together in terms of their ability than B and C, say, whether the easy test or the hard test is administered.

Another observation about Fig. 6.2 is that the relationship between the scores on the two tests is not linear (not a straight line). That is, to map the scores of the hard test onto scores of the easy test, there is not a simple linear transformation such as a constant shift and/or a constant scaling factor. If the relationship between scores on two tests is a straight line, then comparing two students using either test will give the same relative distances between students. Item response modelling provides such a transformation of test scores to make the relationship between scores on two tests a linear one. We clarify this in the latter part of this chapter.

Linking People to Tasks

Another desirable characteristic of measurement is that "meanings" can be given to scores. That is, we would like to know what a student can actually do if the student obtained a score of, say, 55 out of 100 on a test. Therefore if student scores can be linked to the items in some ways, then substantive meanings can be given to scores in terms of the underlying skills or proficiencies. For example, one would like to make statements such as

> Students who obtain 55 out of 100 on this test are likely to be able to carry out two-digit multiplications and solve arithmetic change problems, but they will typically have difficulties with multi-step word problems.

When raw scores (percentages of correct responses) are used to measure students' abilities and item difficulties, it is not immediately obvious how one can link student scores to item scores. For example, Fig. 6.3 shows two scales in relation to a test, one for item difficulty, and one for person ability. The item difficulty scale on the left shows that for the set of word problems in a test, the average percentage of correct responses amounted to 25% for a cohort of students. In contrast, 90% of the students correctly carried out single digit additions.

Next, let us consider the person ability scale which shows students who obtained 90, 70, 50 and 25% correct on the test. The percentages on the two scales are not easily matched in any way. For example, can the students who obtained 70% on the test perform arithmetic with fractions? We cannot make any inference if we do not know what proportions of items are about single digit addition, multi-step arithmetic, or other types. It may be the case that 70% of the items are single-digit addition items, so that the students who obtained 70% correct on the test cannot perform tasks much more difficult than single-digit addition.

Even if we have information on the composition of the test in terms of the number of items for each type of problems, it is still a difficult job to match student scores with tasks. The underlying skills for each test score will need to be

Fig. 6.3 Link raw scores on items and persons

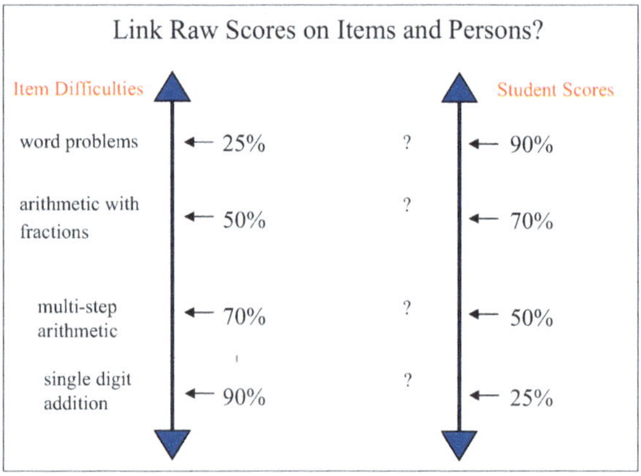

examined, and descriptions written for each test score. For example, we need to examine the common items answered correctly by students with a test score of, say, 25 or 50%, and construct descriptions of skills for these scores. No systematic approach can be taken. When a different test is administered, a new set of descriptions will need to be developed, as there is no simple and direct relationship between student scores and item scores.

Estimating Ability Using Item Response Theory

The problems with using raw scores as discussed above can be solved by using ability estimates from item response theory (IRT) modelling. The main idea of item response theory is to use a mathematical model for predicting the probability of success of a person on an item, depending on the person's "ability" and the item "difficulty". Typically, the probability of success on an item for people with varying ability is plotted as an "item characteristic curve" (ICC). An example ICC is shown in Fig. 6.4, where it takes the shape of an elongated letter "S". The ICC in this example is a logistic function of the form $f(x) = \frac{e^x}{1+e^x}$. An IRT model with a logistic item response function is called the Rasch model (Rasch 1960). This is the "simplest" IRT model in that the item response function is determined by only one parameter (the item difficulty parameter). Chapter 7 further explains this mathematical model, and Chaps. 9 and 10 explain two other models. While many different mathematical functions can be used to model the probability of success of a person on an item, these functions should have three properties. First, the function should be increasing with ability. That is, if the ability is higher, the probability of success should also be higher. Second, the function should take on values of x that ranges between $-\infty$ and ∞. That is, the ability can range from infinitely low to

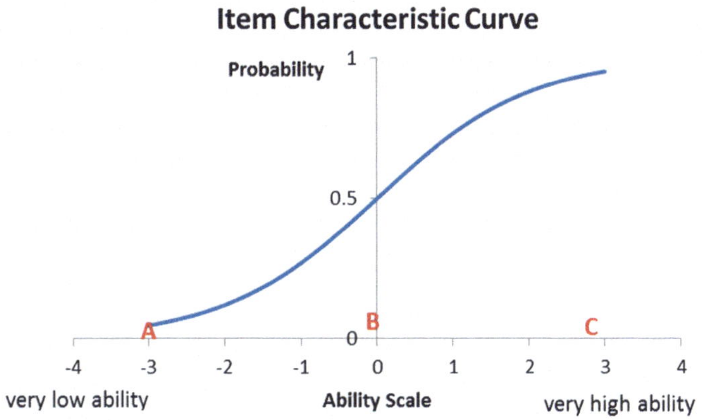

Fig. 6.4 An example of an item characteristic curve

infinitely high, and there is no lower or upper bound. Third, the function should evaluate to a value between 0 and 1, since it is a probability.

Figure 6.4 shows that, for a high achiever (C), the probability of success on this item is close to 1. For a low achiever (A), the probability of success on this item is close to zero. For an average ability student (B), the probability of success is 0.5. The curve through the points A, B and C shows the probability of success on this item at each ability level. This curve is called the item characteristic curve.

Under this model, the item difficulty for an item is defined as the level of ability at which the probability of success on the item is 0.5. In the example given in Fig. 6.4, the ability level of person B (δ) is also the item difficulty of this item, since person B has 50% chance of answering the item correctly. Defined in this way, the notion of item difficulty relates to the difficulty of the task "on average". Obviously for a very able person, the item in Fig. 6.4 is very easy, and for a low ability person, the item is difficult. But the item difficulty (δ) is defined in relation to the ability level of a person who has a 50–50% chance of being successful on the item.

It is important to emphasise that by definition, an item difficulty is a value on the ability scale. That is, both item difficulty and ability are on the same scale. This property is one key difference between item response theory and classical test theory where raw scores are used.

A note should be made about the interpretation of a student's probability of success on an item. How does one explain the stochastic nature of a student's score on an item, given that a student either gets an item right or wrong? Under an IRT model, suppose Tom has a 60% chance (probability) of correctly answering an item. There are at least two ways to think of the probabilistic nature of item responses. The first explanation is to postulate that if the item is repeatedly administered to Tom, then 60% of the time Tom will obtain the correct answer. This explanation is not so attractive, since one may believe that if Tom knows the answer to an item, Tom will (nearly) always know the answer. The second explanation of the probabilistic nature of item responses is to think of groups of items and groups of students. A 60% probability of success for Tom on an item means that about 60% of students at Tom's ability level will answer this item correctly. Further, for a set of items with the same item difficulty, Tom will answer 60% of the items correctly. This second explanation about probabilities for a group of students and for a set of items is our preferred one.

Figure 6.5 shows three item characteristic curves with varying item difficulties. It can be seen that the item with the green ICC (the left-most curve) is the easiest item among the three, while the item with the blue ICC (the right-most curve) is the most difficult. The item difficulties for the three items are denoted by δ_1, δ_2, δ_3, where $\delta_1 < \delta_2 < \delta_3$. For the easiest item, everyone has a higher probability of being successful than for a more difficult item. In this way, the item difficulty parameters, δ_1, δ_2, δ_3, define a clear order of item difficulty.

As item difficulties are defined in relation to ability levels, we can make statements about a person's likelihood of success on an item when item difficulty and ability are known. That is, if we know a student's ability, we can predict how that person is likely to perform on an item (without administering the item to the person)

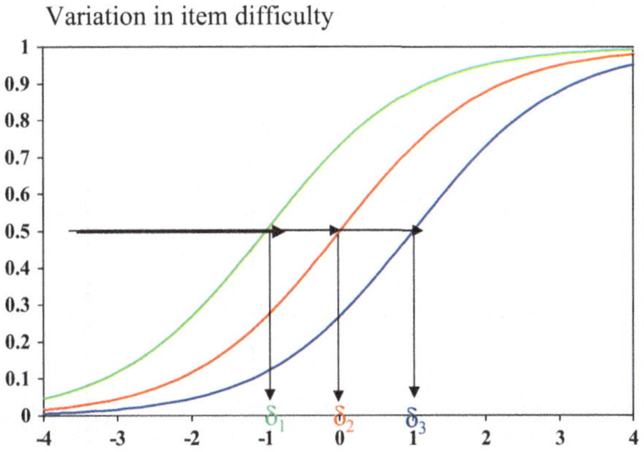

Fig. 6.5 Three ICCs with varying item difficulty

in terms of probability statements, even though we cannot precisely determine whether a student will successfully answer a question. Since there is an underlying mathematical function to model student's item responses, one can make such probability statements about the *chances* of a student obtaining a correct answer. This is an advantage of using a mathematical function to model the probability of success. Of course, the mathematical model should actually reflect the patterns of student response data, or else our predictions will be wrong. This is an assumption underlying the validity of using a particular mathematical model, and this assumption needs to be checked.

Figure 6.6 shows an example of using item characteristic curves to find the probabilities of success on three items if the ability of a person (θ) is known. A vertical line can be drawn in Fig. 6.6 to read off the probability of success on each of the three items for a person with an ability of 0.9.

By defining item difficulty and person ability on the same scale, interpretations of person scores can be easily provided in terms of the task demands. Figure 6.7 shows an example. The person ability scale on the left and the item difficulty scale

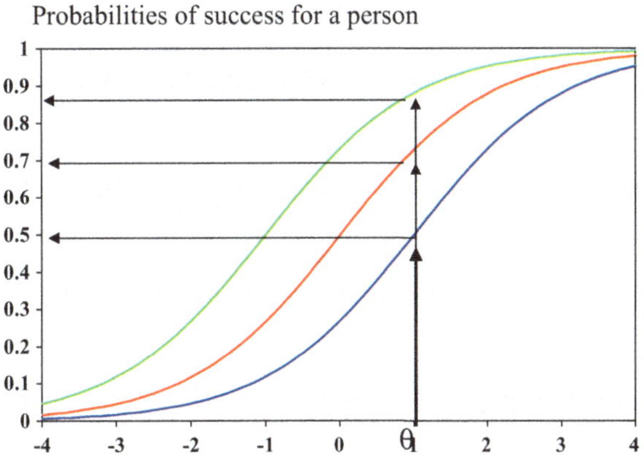

Fig. 6.6 Probabilities of success for a person with ability of 0.9

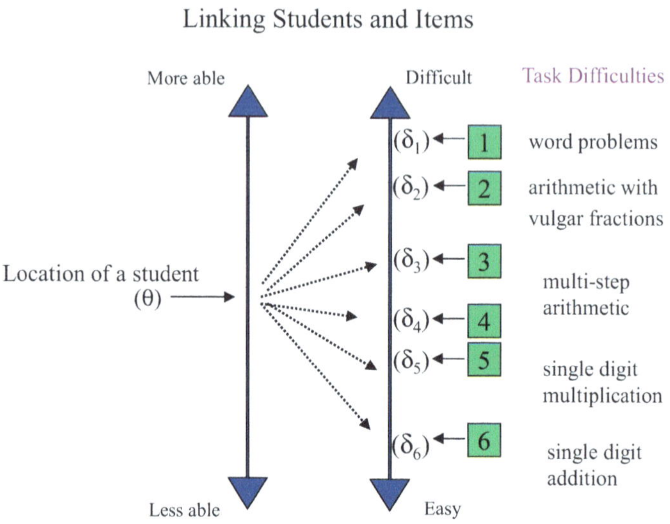

Fig. 6.7 Linking students and items through an IRT scale

on the right are linked through the mathematical function of probability of success. If a student has an ability of θ, one can readily compute this student's chances of success on items 1 to 6, with item difficulties $\delta_1, \delta_2, \ldots, \delta_6$, respectively. As one can describe the underlying skills required to answer each item correctly, one can easily describe a student's level of proficiency once we have located the student on the scale according to his/her ability. For example, a student located at θ in Fig. 6.7 will typically have a 50% chance of successfully carrying out multi-step arithmetic; more than 50% chance of performing single-digit multiplication; and less than 50% chance of performing arithmetic with fractions.

Estimation of Ability Using IRT

To explain about how abilities are estimated under IRT, Figs. 6.8, 6.9, 6.10 and 6.11 present a sequence of illustrated steps.

Given the definition of item difficulties, a student located at θ on the ability scale will typically have a 50% chance of successfully answering an item with difficulty value at θ. Put it another way, for a student with an ability θ, we expect the student to answer about 50% of the items correctly for items with difficulty values around θ. Figure 6.8 shows that about 50% of the items located around the ability of a student are marked correct, and about 50% marked incorrect (a tick shows an item is marked correct, and a cross shows incorrect).

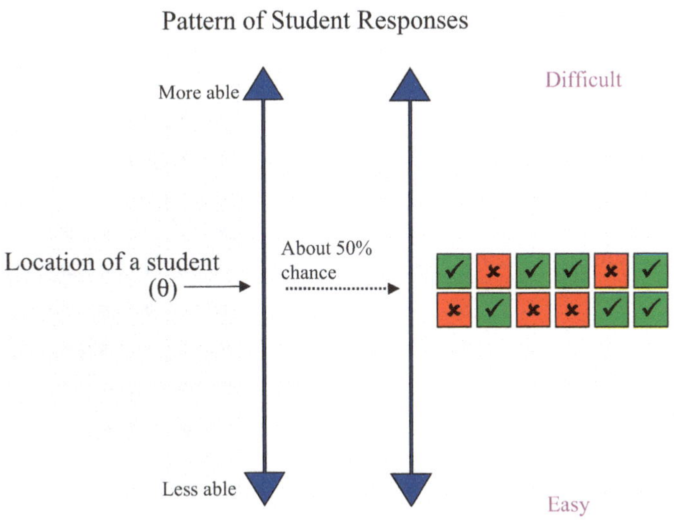

Fig. 6.8 Items at a student's ability level—about 50% correct

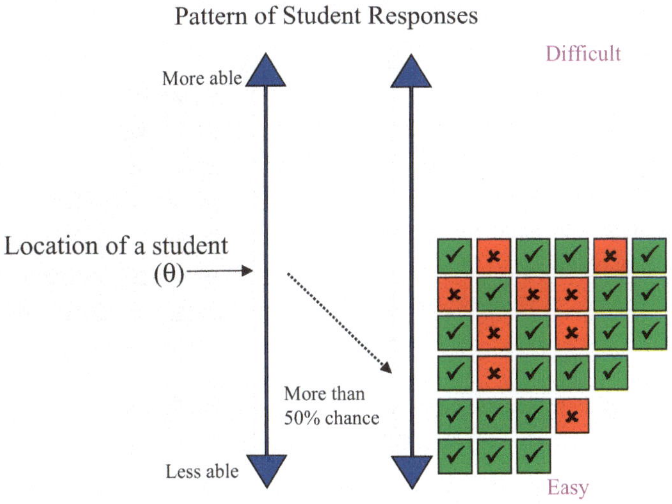

Fig. 6.9 Items located below a student's ability level—more than 50% correct

Next, notice that in Fig. 6.9 there is a block of items located below the ability of a student, and that there are more correct answers than incorrect answers. That is, there are more ticks than crosses for this block of items for this student.

In comparison, Fig. 6.10 shows that the block of items with difficulties higher than the student's ability will typically have more incorrect answers than correct

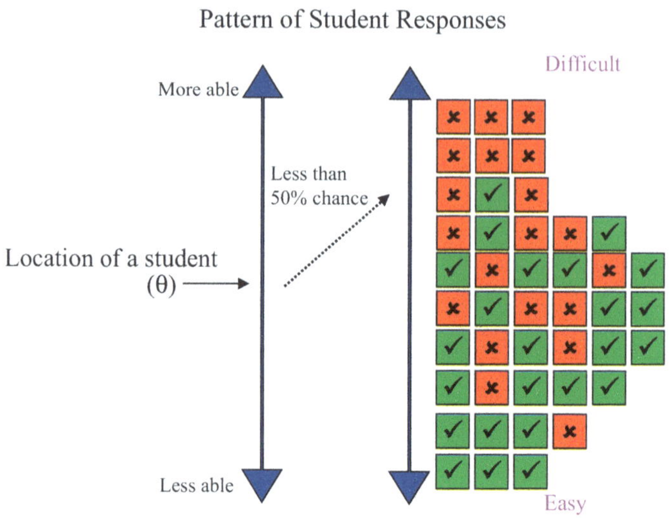

Fig. 6.10 Items located above a student's ability level—less than 50% correct

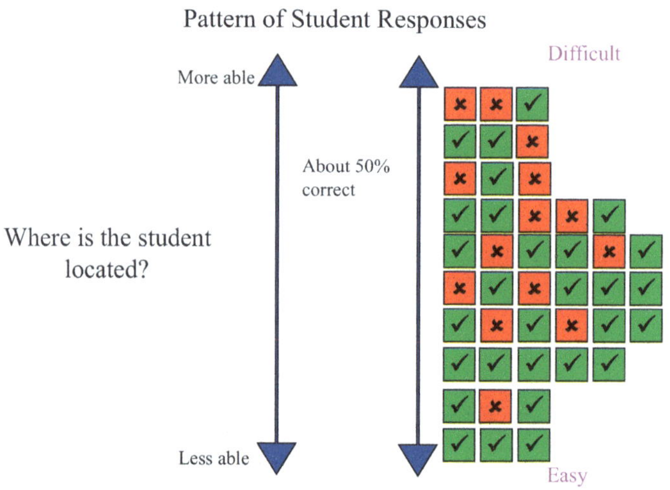

Fig. 6.11 Given item response pattern, find student ability

answers, because the chance of answering these items correctly for the student is less than 50%.

In real-life, what we observe are the item response patterns of a student on a test, that is, the item correct-incorrect patterns on the right-side of Fig. 6.10. We do not know a student's ability. The goal is to use the item response patterns to find the student's ability. What we try to identify is the ability region where about equal numbers of correct and incorrect item responses are located.

Figure 6.11 shows the item response patterns of another student. Can you estimate where the student's ability is located?

Examining the item response patterns in Fig. 6.11, we try to find an ability estimate at which around 50% of the items are correct. In Fig. 6.11, this is close to the top of the scale. Our guestimate is around the third row of items from the top. This could be where we locate the student's ability. Looking for a region where there are about 50% correct answers is the principle of finding student ability estimate in IRT. Of course there is an assumption here that the item difficulties are already known so that we can place item responses at their appropriate places on the scale. What we have illustrated here is the basic principle of estimation in IRT, but there are different estimation methods in IRT, and mathematical procedures are involved rather than the eye-balling procedure as illustrated above.

Invariance of Ability Estimates Under IRT

In an earlier section of this chapter, we discussed about the issues with measurement invariance when raw scores are used. In this section, we illustrate how measurement invariance is achieved under the IRT framework.

Take the item response pattern in Fig. 6.11 as an example, if easy items have *not* been administered, as shown in Fig. 6.12, how would the estimation of ability be affected?

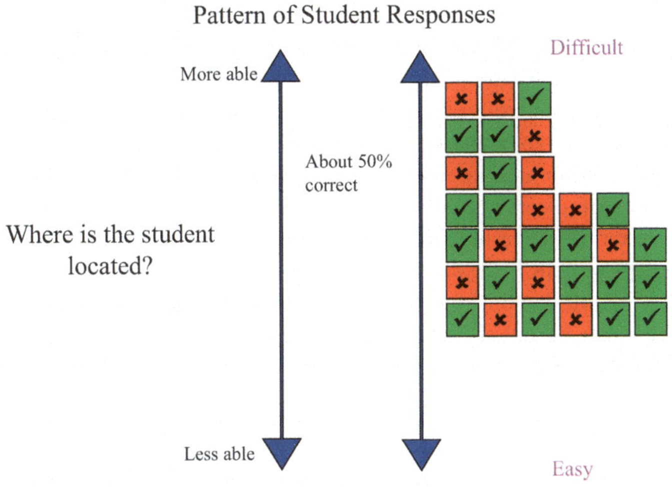

Fig. 6.12 Easy items are not administered

Figure 6.12 is the same as Fig. 6.11 except that the bottom three rows of item responses are removed, indicating that a more difficult test has been administered. If raw scores are used, the student obtained 19/32 = 59% on the harder test as shown in Fig. 6.12, and 29/43 = 67% on the easier test as shown in Fig. 6.11. To determine ability estimate under the IRT principle for item responses in Fig. 6.12, we identify a region where about 50% of the items are correct. This region is actually not changed whether the bottom three rows of items are present or not. Still, in the region near the top third row, about 50% of the items are correct. From this example, it can be seen that whether an easy test or a hard test is administered, the location of the ability estimate is unaffected. This is because IRT uses a probability model rather than raw scores or percentages of items correct in determining ability estimates. In this way, measurement invariance is attained.

Computer Adaptive Tests Using IRT

The example above illustrates that the principles of IRT enable the estimation of student abilities even when students take different sets of items. Consequently, IRT is particularly useful in computer adaptive testing where items are selected for individual students depending on the ability level of each student. A high level student typically gets more difficult questions, and a low level student gets easier questions. But IRT can provide ability estimates that are comparable across students who take the computerized test. Refer to Chap. 7 for more information.

Summary

This chapter introduces the desirable properties of an ideal measurement. The properties include measurement invariance and interpretability of measures. Measurement invariance refers to the invariance of the placement of students on the ability scale irrespective of the instruments administered, provided, of course, that the instruments all tap into the same construct. Interpretability of measures refers to the attachment of meanings to measures, so that measurement scores can be interpreted in terms of the underlying skills that the students can perform.

The principles of IRT are discussed conceptually and contrasted with classical test theory (CTT) to demonstrate how IRT overcomes some shortcomings of CTT. It should however be noted that the differences between IRT and CTT are conceptual. In practice, the differences between IRT and CTT are not so great when the difficulties of instruments are well designed to match with students' abilities. When test length is long, CTT provides sufficiently good measures with raw scores that

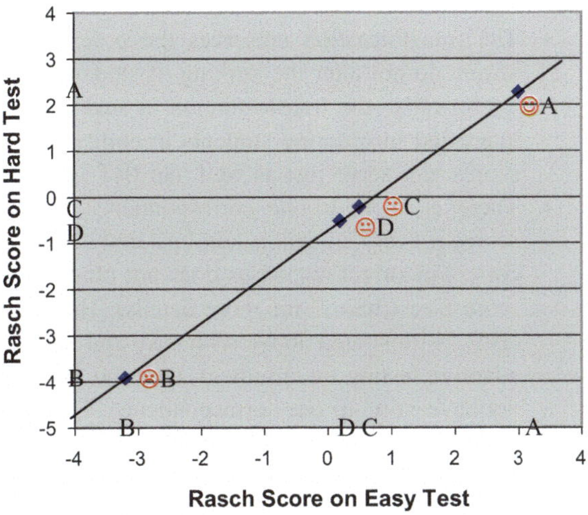

Fig. 6.13 A plot of student IRT Rasch scores on an easy test and a hard test

can be regarded as close to interval measurement. Nevertheless, the good measurement properties of IRT can be applied to build better assessment systems such as the use of rotated test booklets and computer adaptive testing. CTT is quite limited to the analysis of single tests.

Additional Notes

IRT Viewed as a Transformation of Raw Scores

The Rasch model is a particular IRT model. The Rasch model can be viewed as applying a transformation to the raw scores so that distances between the locations of two students can be preserved independent of the particular items administered. The curved line in Fig. 6.2 will be "straightened" through this transformation. Figure 6.13 shows an example of this transformation. Note that the distance between A and C on the easy test (horizontal axis) is the same as the distance between A and C on the hard test (vertical axis).

A crude transformation from raw test score to an IRT ability score is

$$\theta = \log\left(\frac{p}{1-p}\right)$$

where θ is IRT ability and p is the raw score in percentage (e.g., $p = 0.8$, if the raw score is 80% correct on the test).

A number of points can be made about IRT (Rasch) transformation of raw scores:

- The transformation preserves the order of raw scores. That is, Rasch scores do not alter the ranking of students according to their raw scores. Technically, the transformation is said to be monotonic. If one is only interested in ordering students in ability, or items in difficulty, then raw scores will serve just as well. No IRT is needed.
- There is a one-to-one correspondence between raw scores and Rasch scores if every student is administered the *same* test. That is the pattern of correct/incorrect responses does not play a role in determining the Rasch score (see Chap. 7 for more details). However, if students take *different* tests, as illustrated above with easy and hard tests, and within a computer adaptive testing environment, then the raw scores and Rasch scores will not have a one-to-one correspondence. The Rasch scores will take the item difficulties of the overall test into account.
- When students take the same test, the correlation between raw score and Rasch score will be close to 1, as a result of the property of the Rasch model. Occasionally, one sees researchers plotting Rasch scores against raw scores. The high correlation between these two scores has sometimes been taken as indications of good fit of the data to the model. This is a misconception. Actually, even if data mis-fit the model, the correlation between Rasch scores and raw scores will still be close to one.

How About Other Transformations of Raw Scores, for Example, Standardised Score (Z-Score) and Percentile Ranks? Do They Preserve "Distances" Between People?

Using classical test theory approach, raw scores are sometimes transformed to z-scores or percentile ranks. For z-scores, a transformation is applied to make the mean of the raw scores equal to zero, and the standard deviation equal to 1. This transformation is linear, so the relative distance between two points will be the same whether raw scores or z-scores are used. For example, if A and C are further apart than C and B in raw scores, then the z-scores will also reflect the same relative difference. Consequently, z-scores suffer from the same problem as raw scores. That is, z-scores on an easy test and a hard test will not necessarily preserve the same relative distances between students.

Transforming raw scores to percentile ranks will solve the problem of producing differing distances between two people on two different tests. This is because percentile ranks have relinquished the actual distances between students, and turned the scores to ranks (ordering) only. So, on the one hand, the percentile ranks of people on two different tests may indeed be the same, on the other hand, we have lost the actual distances between students. Raw scores, while not quite providing an interval scale, offer more information than just ordinal scales.

Hands-on Practices

Task 1

Use simulation to generate raw scores for students on an easy test and a hard test.

Q1. Plot the two test scores on a graph
Q2. Apply a logistic transformation to the raw scores as follows:

Step 1: Compute percentage correct from the raw scores (raw score divided by possible maximum score). Let p denote percentage correct.
Step 2: Compute transformed score by applying transformation, log(p/(1 − p)), where log is the natural logarithm. The ratio, p/(1 − p), is referred to as an "odds". The results from the transformation of log(p/(1 − p)) are said to be in the "log of odds unit" (abbreviated as "logit")
Step 3: Plot the two transformed scores on a graph

Discuss the shapes of the two graphs in terms of measurement invariance. Which graph is closer to a straight line?

Note: This hands-on practice is to demonstrate IRT as viewed as a transformation of the raw scores. However, the actual mathematical modelling of IRT is at the individual item and individual person level, not at the test score level. In IRT software programs, often logistic transformations applied to the test scores or to item scores (percentage of students getting an item right), as shown in this hands-on practice, are used to provide initial values of person and item parameters.

Task 2

Investigate the relationship between raw scores and transformed logit scores. For example, if a test has a maximum score of 30, plot raw scores (between 0 and 30) against transformed scores. What are your observations in terms of the distances between raw scores and between logit scores? Is the relationship between raw scores and logit scores a linear one? If not, is there a range between which the relationship is approximately linear?

Discussion Points

(1) For what purposes of measurement would raw scores be sufficient? For what purposes of measurement should IRT be applied?

(2) Based on the presentation in Chaps. 5 and 6, what do you think are the differences between classical test theory and item response theory?

(3) The illustration of the principles for estimating ability (as shown from Figs. 6.8, 6.9, 6.10 and 6.11) relies on a response pattern that shows more items correct for easy items, and fewer items correct for difficult items. In this way one can identify the region where there are about equal numbers of correct and incorrect items. What happens if there is no clear pattern of item responses, such as a random scattering of incorrect items over the low to high scale, so that there is no clear region where the student's ability might be?

Exercises

Q1. As percentages, raw scores have a minimum of 0 and a maximum of 100. What is the minimum and maximum of logits? (logit is defined as in the Hands-on Practice section).

Q2. When percentage (p) is 50%, what is the value of the transformed logit?

Q3. Consider two raw scores expressed in percentages, p1 and p2, where p2 is greater than p1. Let t1 and t2 denote the transformed logit scores of p1 and p2 respectively. Which of the following option(s) do you think are appropriate in relation to the relative magnitude of t1 and t2?

t1 is greater than t2
t2 is greater than t1
One cannot say which is larger, as it depends on whether t1 and t2 are positive or negative
One cannot say which is larger, as it depends on whether p1 and p2 are below or above average

Q4. The following shows the response pattern of a student. Can you estimate the student's ability?

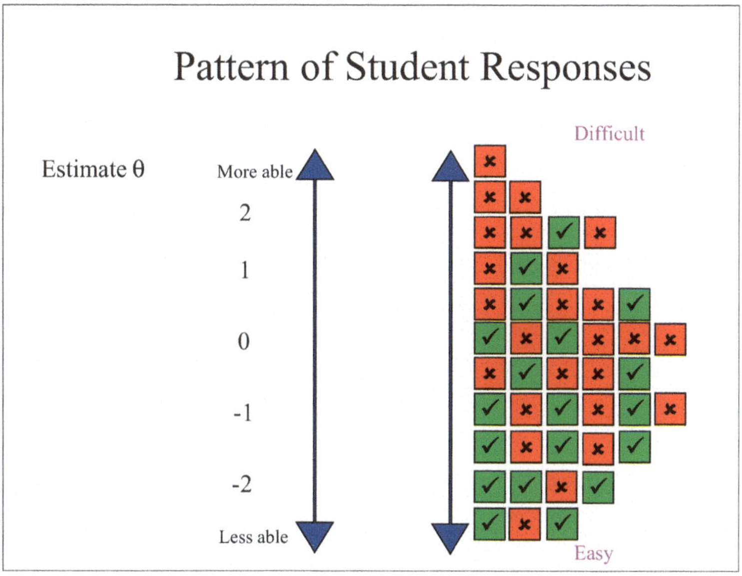

Reference

Rasch G (1960) Probabilistic models for some intelligence and attainment tests. Danish Institute for Educational Research, Copenhagen

Further Reading

Bond TG, Fox CM (2007) Applying the Rasch model: fundamental measurement in the human sciences, 2nd edn. Lawrence Erlbaum Associates, Mahwah, NJ
Engelhard G (2013) Invariant measurement: using Rasch models in the social, behavioural, and health sciences. Routledge, New York, NY
Wilson M (2005) Constructing measures: an item response modeling approach. Lawrence Erlbaum Associates, Mahwah, NJ
Wright BD, Masters GN (1982) Rating scale analysis. Mesa Press, Chicago
Wright BD, Stone MH (1999) Measurement essentials. Wide Range, Inc., Wilmington, DE. http://www.rasch.org/measess/me-all.pdf

Chapter 7
Rasch Model (The Dichotomous Case)

Introduction

There are many different IRT models. The simplest model specification is the dichotomous Rasch model. The word "dichotomous" refers to the case where each item is scored as correct or incorrect (0 or 1).

The Rasch Model

Item response models typically apply a mathematical function to model the probability of a student's response to an item. The probability is a function of the student's "ability" level. The graph of the probability function is usually known as item characteristic curve (ICC), which typically has an "S" shape as shown in Fig. 7.1.

In the case of the Rasch model (1960), the mathematical function of the item characteristic curve for a dichotomous item is typically given by

$$p = P(X = 1) = \frac{\exp(\theta - \delta)}{1 + \exp(\theta - \delta)} \tag{7.1}$$

where X is a random variable indicating success or failure on the item, with $X = 1$ indicates success (or a correct response) on the item, and $X = 0$ indicates failure (or an incorrect response) on the item.

θ is a person-parameter denoting the person's ability on the latent variable scale, and δ is an item-parameter, generally called the item difficulty, on the same latent variable scale. The Rasch model is sometimes called the one parameter model (1PL), since the function in Eq. (7.1), when expressed as a function of the ability θ, has one parameter, namely, the delta (δ) parameter.

M. Wu et al., *Educational Measurement for Applied Researchers*,
DOI 10.1007/978-981-10-3302-5_7

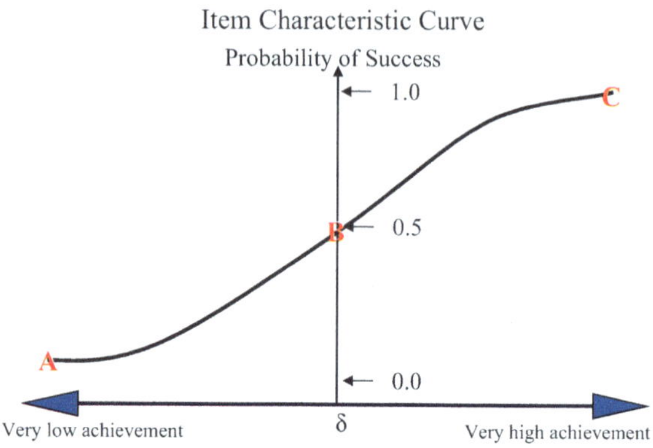

Fig. 7.1 An example item characteristic curve

Equation (7.1) shows that the probability of success on an item is a function of the difference between a person's ability and the item difficulty. When the ability equals the item difficulty, the probability of success is 0.5.

By re-arranging terms and then taking logarithm on both sides of Eq. (7.1), it is easy to demonstrate that

$$\log\left(\frac{p}{1-p}\right) = \theta - \delta \tag{7.2}$$

Equation (7.2) shows that $\theta - \delta$, the distance between a person's ability and the item difficulty, is expressed as the logarithm of the *odds* of success of the person on the item. The term odds is the ratio of the probability of success over the probability of failure. As a result, the measurement unit of the scale for ability and item difficulty is generally known as "logit", a contraction of "log of odds unit".

Moreover, if one interprets p as the percentage of items with difficulty δ answered correctly by students with ability θ (see Chap. 6 for the interpretations of p), one can think of $\log\left(\frac{p}{1-p}\right)$ as a transformation of p (percentage correct) and this transformed score is on the logit scale ($=\theta - \delta$). In this way, the ability score in logits can be viewed as a transformation of the percentage correct, in much the same way as other scaled scores which are transformations of the raw scores, as discussed in Chap. 6. In fact, in some IRT software programs, the initial values for item difficulty estimates are often set as $\log\left(\frac{p}{1-p}\right)$ where p is the percentage of students who obtained the correct answer on an item. Similarly, $\log\left(\frac{p}{1-p}\right)$ can be used as initial values for person ability estimates, where p is a student's test score expressed as the percentage of correctly answered items.

Additional Notes

Many IRT models use the logistic item response function although the logistic function is not the only function that can be used (e.g., see Embretson and Reise 2000; van der Linden and Hambleton 1997; Thissen and Steinberg 2009). The choice of the item response function is not simply for mathematical convenience. There are theoretical reasons why item response data may follow the logistic model (e.g., Rasch 1960; Wright 1977). It has also been shown empirically that item response data do generally fit the logistic model (e.g., Thissen and Wainer 2001). In addition to logistic functions, the normal ogive function has also been used (Lord and Novick 1968; Samejima 1977). In general, the normal ogive model can be approximated by the logistic item response model (Birnbaum 1968). See Hands-on Practices Task 2 for more information.

Properties of the Rasch Model

Specific Objectivity

Rasch (1977) pointed out that the model specified by Eq. (7.1) has a special property called *specific objectivity*. The principle of specific objectivity is that comparisons between two objects must be free from the conditions under which the comparisons are made. For example, the comparison between two persons should not be influenced by the specific items used for the comparison. To demonstrate this principle, consider the log odds for two persons with abilities θ_1 and θ_2 on an item with difficulty δ. Let p_1 be the probability of success of person 1 on the item, and p_2 be the probability of success of person 2 on the item. Substituting into Eq. (7.1), we have

$$\log\left(\frac{p_1}{1-p_1}\right) = \theta_1 - \delta$$

$$\log\left(\frac{p_2}{1-p_2}\right) = \theta_2 - \delta \tag{7.3}$$

The difference between the log odds for the two persons is given by

$$\log\left(\frac{p_1}{1-p_1}\right) - \log\left(\frac{p_2}{1-p_2}\right) = \theta_1 - \delta - (\theta_2 - \delta) = \theta_1 - \theta_2 \tag{7.4}$$

Equation (7.4) shows that the difference between the log odds for two persons depends only on the ability parameters and not on the item parameter. That is, irrespective of which items are used to compare two persons, the difference between the log odds for the two persons is the same.

Similarly, it can be demonstrated that the comparison between two items is *person-free*. That is, the difference between the log odds for two items is the same regardless of which persons took the two items.

Some psychometricians regard this *sample-free* property of a measurement model as most important for constructing sound measurements, because statements can be made about relative item difficulties without reference to specific persons, and similarly statements can be made about relative proficiencies of people without reference to specific items. This item- and person-invariance property may not hold for some IRT models, such as the two-parameter and three-parameter IRT models.

Indeterminacy of an Absolute Location of Ability

Equation (7.1) shows that the probability of success of a person on an item depends on the difference between ability and item difficulty, $\theta - \delta$. If one adds a constant to ability θ, and then adds the same constant to item difficulty δ, the difference between ability and item difficulty, $\theta - \delta$, will remain the same, so that the probability of success will remain the same. Consequently, the logit scale does not determine an absolute location of ability and item difficulty. The logit scale only determines relative differences between abilities, between item difficulties, and between ability and item difficulty. This means that, in scaling a set of items to estimate item difficulties and abilities, one can choose an arbitrary origin for the logit scale, and that the resulting estimates are subject to a location shift without changing the fit to the model.

To emphasise further this indeterminacy of the absolute location of ability and item difficulty estimates, one must not associate any interpretation to the logit value without making some reference to the nature of the origin of the scale however it was set. For example, if an item has a difficulty value of 1.2 logits from one scaling, and a different item has a difficulty value of 1.5 logits from another scaling, one cannot make any inference about the relative difficulties of the two items without examining how the two scalings were performed in terms of setting the origins of the scales and how the two scales are linked. The readers are referred to Chap. 12 on equating for more information on this point.

This indeterminacy of the origin of a scale is not specific to IRT models. It is a matter of fact that when students take a test, if the test scores are high, we will not know whether it is because students are able, or if the test items are easy, without making additional assumptions about the students and/or the test. That is, student abilities and item difficulties are confounded when test scores are obtained. Understanding this important point will help greatly with more complex applications of test analysis, such as making comparisons between groups of students, or

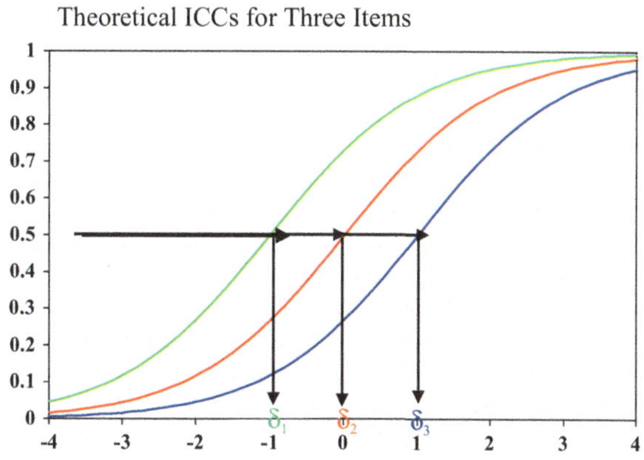

Fig. 7.2 Three ICCs with varying item difficulties but the same discrimination

tracking students over time, or comparing scores across a set of tests. In short, item difficulty for an item is always with reference to a group of people the item is administered to, or with reference to a group of items. Consequently, we need to clarify that the property of specific objectivity is in the context of relative measures between items, between respondents, and between item and respondent, and NOT in the individual measures. The term "sample-free" as found in many literature can be misleading if it is not interpreted appropriately.

Equal Discrimination

Under the Rasch model, the *theoretical* item characteristic curves for a set of items in a test are all *parallel*, in the sense that they do not cross, and that they all have the same shape except for a location shift, as shown in Fig. 7.2. This property is known as *equal discrimination* or *equal slope parameter*. That is, each item provides the same *discriminating power* in separating respondents by their levels on latent trait.

Indeterminacy of an Absolute Discrimination or Scale Factor

Figure 7.3 shows three ICCs corresponding to three items. The ICC for the first item (blue dotted curve) is rather flat, while the ICC for the third item (green solid line) is quite steep. Of course, for each ICC, the slope varies along the ability range. For example, for Item 3, the curve is steep over the middle ability range, and flat over other parts of the ability range. In the following discussion, we refer to the slope of an ICC as the slope at around the item difficulty value. When the slope of an ICC is flat, such as for Item 1, students at low ability levels have similar probabilities of answering the item correctly as for students at high ability levels.

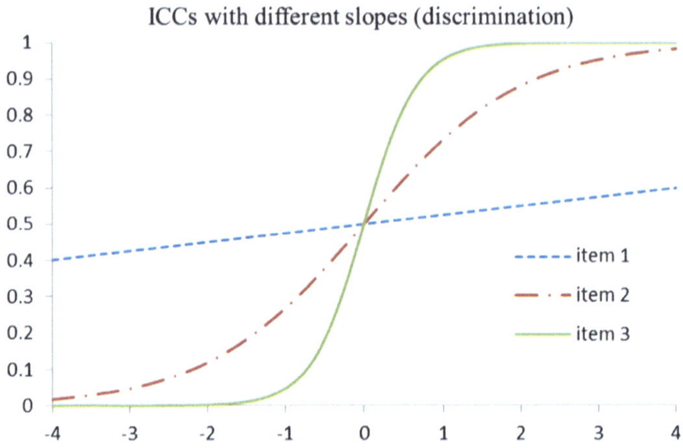

Fig. 7.3 Three ICCs with different discrimination but the same difficulty

Consequently, the item does not provide much power in discriminating students of varying abilities. In the extreme case where the ICC is a horizontal line, then the item cannot distinguish between low from high ability students at all. In summary, the *slope* of an item characteristic curve shows an item's discrimination power.

In terms of the mathematical formulation of the slope parameter, Eq. (7.5) shows a model that takes the slope into account.

$$p = P(X = 1) = \frac{\exp(a(\theta - \delta))}{1 + \exp(a(\theta - \delta))} \tag{7.5}$$

The value of a in Eq. (7.5) determines the slope of the ICC. In Fig. 7.3, a takes the values of 0.1, 1 and 3 for Item 1, Item 2 and Item 3 respectively. Note that Eq. (7.5) is not a Rasch model if the value of a differs for different items, since the Rasch model assumes all items in a test have the same discrimination and, as a convention, the value of a is set to 1. Also note that since all items have the same slope under the Rasch model, the ICCs do not cross each other. Equation (7.5) shows the two-parameter IRT model (2PL) when each item has a different value of a. A detailed discussion of 2PL models can be found in Chap. 10.

While the Rasch model stipulates that all items in a test have the same "discrimination" (or the same "slope"), the Rasch model does not specify an absolute value for the discrimination parameter. The setting of a to 1 is a convention only. We can set a to any constant. Provided that all items have the same a we have the Rasch model. (See Hands-on Practices Task 2 for more information.) Since $a(\frac{\theta}{a} - \frac{\delta}{a}) = (\theta - \delta)$, the ability scale can have any scale factor. For example, most part of the range of ability estimates could be between −3 and 3, or between −300 and 300, or between 100 and 800. The model will fit equally well by multiplying/dividing a scale constant to all abilities and item difficulties, and setting an arbitrary origin. Chapter 10 shows how the a parameters can be transformed by a scale factor.

An additional note: For this reason we do not hold the view that if a is set to 1.7 it is the 1PL model, and if a is set to 1.0, it is the Rasch model. In both cases, the model is the Rasch model, just with a different scale factor. The scale factor a can be any number and it is still the same model because of the indeterminacy of the scale factor. See Hands-on Task 2 for the reason for setting a to 1.7.

Different Discrimination Between Item Sets

As an example to illustrate relative discrimination between items and the setting of the a parameters, Fig. 7.4 shows two sets of items with different discriminating power when the two sets of items are administered together to the same group of people. While items within each set have the same "slope", Set 2 items (right-side graph) are more discriminating than Set 1 items (left-side graph).

When each set of items is scaled using the Rasch model in two separate scaling runs, the slope parameter of the item characteristic curve is set to a "1" *as a convention* (i.e., the value of a in Eq. (7.5) is set to 1) in each run, so that the two sets of items appear to have the same slope pictorially (Fig. 7.5). However, students taking Set 2 items will have ability estimates that are more spread out. (See the change in the scale of the horizontal axes of the ICCs from Figs. 7.4 to 7.5.) That is, the variance of the ability distribution using Set 2 items will be larger than the variance of the ability distribution when Set 1 items are used. Consequently, the reliability of a test using Set 2 items will be higher, remembering that reliability shows the extent to which a test can separate students (refer to Chap. 5). To demonstrate the shrinking and expansion of the scale, imagine the graphs in Fig. 7.4 are re-sized using Windows re-size tool (⇔). To make the ICCs steeper in the left-side graph, the sides of the window need to be brought towards each other. To make the ICCs flatter in the right-side graph, the sides of

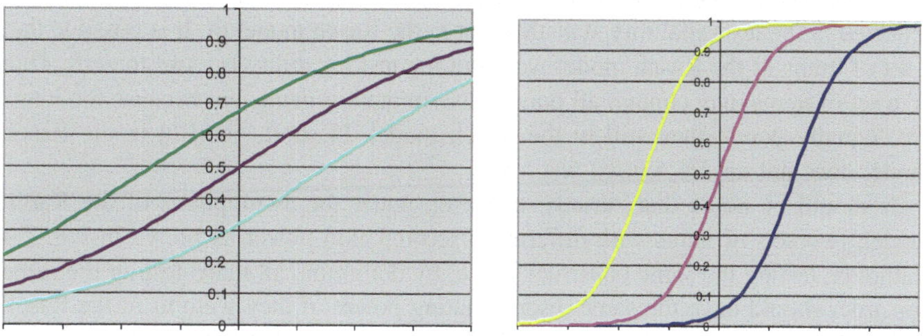

Fig. 7.4 Two sets of items with different discriminating power

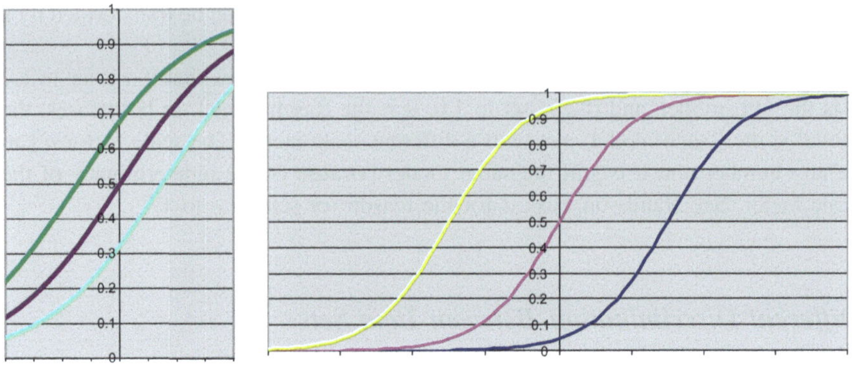

Fig. 7.5 Two sets of items, after separate Rasch scaling

the window need to be pulled further apart. In this way, the scale is shrunken and expanded respectively. More specifically, the slope parameter is directly related to the scale factor of abilities.

Irrespective of the scale, since each set of items show parallel ICCs, Set 1 items fit the Rasch model equally well as the fit of Set 2 items to the Rasch model. But if the two sets are combined into one test, the items will show misfit to the Rasch model.

Length of a Logit

The above results show that the length of one unit "logit" does not have an absolute meaning. A group of students can be close together in terms of their abilities estimated from one calibration of a test, and be further apart from the calibration of another test. How far apart a group of people are spread on the ability scale depends on the discriminating power of the items used. Clearly, less discriminating items have less power in separating respondents in terms of their abilities, even when the items fit the Rasch model well. The overall discriminating power of a set of items is reflected in the test reliability statistics, not in the Rasch model fit. It is possible that a set of items fit the Rasch model well, but the test reliability is close to zero. That is, a set of items may contain all poorly discriminating items, but because the items are "equally poor", they still fit the Rasch model. In short, good fit to the Rasch model does not ensure a good test.

It should be noted that, strictly speaking, under the assumptions of the Rasch model, two sets of items with differing discrimination power as shown in Fig. 7.4 cannot be testing the same construct, since, by definition, all items testing the same construct should have the same discriminating power, if they were to fit the Rasch model.

However, in practice, the notion of equal discriminating is only approximate, and items in a test often have varying discriminating power. For example, open-ended items are often more discriminating than multiple-choice items when both are used to test the same construct. It would seem reasonable to assume that items could tap into the same construct even if they have different discriminating power. Under this circumstance, different items have different amount of "noise" in measuring a construct, analogous to a scenario of target shooting at 50 paces or at 100 paces. Various relaxations of the assumption of the Rasch model regarding equal discrimination have been made, from allowing different discrimination parameters for groups of items (Humphry and Andrich 2008), to having integer item category scores (similar to awarding partial credit scoring as presented in Chap. 9) (Verhelst and Glas 1995), to the 2PL model where every item has its discrimination parameter that is not restricted to integer values.

In any case, we should be aware of the implications of issues regarding the "length" of a logit, particularly when we select items for equating purposes (see Chap. 12 for equating tests).

Building Learning Progressions Using the Rasch Model

Under the Rasch model, the item characteristic curves (ICC) are "parallel", as shown in Fig. 7.2. Having parallel ICCs enables one to clearly discuss the notion of "item difficulty" across items, since it can be seen from Fig. 7.2 that the *relative* difficulties of the three items stay the same for students at all ability levels. Item 3 (the ICC on the right) is the most difficult item while Item 1 (the ICC on the left) is the easiest item out of the three *for all ability values*. In contrast, if three items have ICC such as the ones shown in Fig. 7.3 where the ICCs are not parallel, it becomes problematic to define "item difficulty order", since Item 3 is the easiest item for high ability students, but it is the most difficult item for low ability students.

Researchers have frequently used item response modelling to build learning progressions (sometimes known as proficiency scales), typically through the use of item-person maps (also known as Wright maps (Wilson 2011)) as shown in Fig. 7.6.

In an item-person map, the ability scale is displayed vertically. The items are placed along the ability scale according to their difficulty values. In Fig. 7.6, it can be seen that Item 17 is the most difficult item (at the top of the scale), while Item 42 is the easiest item (at the bottom of the scale). Summary statements of skills are written along the ability scale based on the locations of test items placed on the scale. These summary statements are descriptions for a learning progression. As learning progressions typically apply to a population, the placement of the items in difficulty order should be valid for that population. In the example in Fig. 7.6, Items 17, 34 and 11 are in decreasing difficulty order, not just for students near the top of the ability scale, but also for all students in this population. In contrast, if items have ICCs that are not parallel, then the ordering of items by difficulty will not be the

Item-person map

```
  3            |                                   |
               |                                   |
               |                                   |
               |17                                 |
               |                                   |
               |34                                 |
               |                                   |
               |                                   |
  2           X|                                   |
              X|                                   |
             XX|                                   |
             XX|11                                 |
              X|                                   |
             XX|37                                 |
            XXX|15                                 |
            XXX|                                   |
           XXXX|                                   |
  1        XXXX|16 19 23                           |
        XXXXXXX|                                   |
         XXXXXX|3 21 48                            |
         XXXXXX|18 44                              |
       XXXXXXXX|                                   |
      XXXXXXXXX|9 10                               |
       XXXXXXXX|35 39 45                           |
     XXXXXXXXXX|22 29                              |
  0   XXXXXXXXX|                                   |
       XXXXXXXX|12                                 |
        XXXXXXX|46                                 |
       XXXXXXXX|4 33 43 47                         |
       XXXXXXXX|14 20 28 38                        |
     XXXXXXXXXX|5                                  |
       XXXXXXXX|                                   |
         XXXXXX|7 24 40                            |
           XXXX|13                                 |
 -1       XXXXX|26 31                              |
         XXXXXX|27                                 |
            XXX|                                   |
             XX|25                                 |
             XX|36                                 |
             XX|6                                  |
              X|1 8 30 32                          |
              X|                                   |
              X|2                                  |
 -2           X|                                   |
               |41                                 |
               |                                   |
               |                                   |
               |                                   |
               |                                   |
               |                                   |
               |                                   |
 -3            |42                                 |
               |                                   |
               |                                   |
====================================================
```

Fig. 7.6 Item-person map

same for students at different ability levels. So it becomes problematic to construct a learning progression for a population, since the learning progression will vary depending on the ability level of a student.

In summary, under the Rasch model, item difficulty order is well defined across the whole ability range. For this reason, some measurement specialists consider the Rasch model to have better measurement properties. In practice, however, there are always degrees of misfit of items and uncertainty in skills audits of items, the ordering of items by difficulty will not necessarily be too different when other IRT models are used. Nevertheless, the Rasch model has particular desirable theoretical properties as compared to some other IRT models when it comes to constructing learning progressions.

Additional Notes on Item-Person Map

Typically, an item-person map is drawn with the respondents located on the left-side of the scale according to ability measures (see the symbol of "x" in Fig. 7.6) and with items located on the right-side according to item difficulty measures (labelled by item numbers). This is possible since one feature of IRT is that person ability and item difficulty are calibrated on the same scale. By definition of item difficulty, students with ability equal to the difficulty of an item will have 50% chance of being successful on the item. Using Fig. 7.6 as an example, students with ability 1 will have 50% chance of being successful on Items 16, 19 and 23.

The following is a list of *mis*-interpretations of the item-person map.

1. Items located above the distribution of respondents are not answered correctly by any respondents. For example, Items 17 and 34 in Fig. 7.6 are located higher than all respondents and therefore no one got these items right. This is an incorrect interpretation as respondents located close to 2 on the logit scale will have slightly less than 50% chance of obtaining the correct answer on these two items. In fact, all respondents have a positive probability of getting these items right, although some of these probabilities could be small.
2. Similarly, a second mis-interpretation is that items located below the distribution of respondents are answered correctly by every respondent, for example, Items 41 and 42. This is an incorrect interpretation as respondents located just above −2 on the logit scale have just slightly over 50% chance of obtaining the right answer on these two items.
3. The expected raw score for a respondent can be obtained by counting the number of items below the ability of the respondents. For example, in Fig. 7.6, for respondents with ability zero, there are 28 items located below zero, so the expected test score for these respondents is 28. This is a mis-interpretation. The expected score of a respondent at an ability level depends on the actually item difficulty measures of all items. Take a simple example. Test 1 has four items all with difficulty values of −1.

Test 2 has four items all with difficulty values of −2. A respondent located at zero ability will be above the four items for both tests. But the expected score for this respondent on Test 1 will be lower than his expected score on Test 2 because the four items in Test 1 are more difficult than the four items in Test 2.

Additional Notes on Response Probability (RP) in relation to Item-Person Map

On the item-person map shown in Fig. 7.6, items are located by their difficulty and persons are located by their ability. By definition of item difficulty, students with ability equal to the difficulty of an item will have 50% chance of being successful on the item. That is, when a person is located next to an item on the item-person map, the person has 50% chance of answering the item correctly. When items are matched to a person to describe what the person can do in relation to the items, it is often felt that, with 50% chance of answering an item correctly, the probability is too low to warrant a statement that the person "can" answer the item. Consequently, the stakes are often raised by increasing the response probability to higher than 50% before stating that a person can perform the task demand of an item. There are variations in deciding on the appropriate probability level at which one can make the statement that a student "can do" an item. Sometimes 65% is used, sometimes 75–85% are used. That is, it is a somewhat arbitrary decision. The probability deemed as appropriate to match a person to the items is sometimes called RP (response probability). Once an RP is decided, the item-person map is often adjusted by shifting the items up or shifting the persons down the map, so that when a person is located next to an item, the probability of success is equal to RP.

For the Rasch model, given that all items have parallel ICCs, it is easy to compute an adjustment to the ability (or item difficulty) for a given RP. Since $\log\left(\frac{p}{1-p}\right) = \theta - \delta$ (see Eq. (7.2)), $\log\left(\frac{RP}{1-RP}\right)$ gives the difference between ability and item difficult to achieve a probability of RP. For example, if $RP = 0.75$, then $\log\left(\frac{0.75}{0.25}\right) = 1.1$. So to align persons to items on the item-person map where $RP = 0.75$, either add 1.1 to all item difficulties, or subtract 1.1 from all person abilities.

Raw Scores as Sufficient Statistics

Under the Rasch model, there is a one-to-one correspondence between a person's estimated ability in logits and his/her raw score on the test. That is, students with the same raw score will be given the same ability estimate in logits, irrespective of which items they answer correctly. Statistically, the raw scores are termed "sufficient statistics" for ability estimates. An explanation for this may be construed as follows: if all items have the same discriminating power, then each item should contribute the same amount of information on ability estimation so the items should have the same weight in determining ability, regardless of whether they are easy or difficult items.

So if you have found that the correlation is close to 1 between the raw scores and Rasch ability estimates in a test, be aware that the Rasch model dictates this relationship. It does not show anything about how well your items worked.

However, if two persons were administered different sets of items, raw scores will no longer be sufficient statistics for their ability estimates. This occurs when rotated test booklets are used, where different sets of items are placed in different booklets, or when computer adaptive tests are administered where each student takes a different set of items. It is also the case when items with missing responses are treated as if the items were not-administered, so that people with different missing response patterns are regarded as being administered different tests. Under these circumstances, the raw score will no longer be sufficient statistic for the ability estimate.

How Different Is IRT from CTT?

Given that item difficulty estimates from the Rasch model has a one-to-one correspondence with item scores (percentages of students obtaining the correct answer), and that ability estimates from the Rasch model has a one-to-one relationship with students' test scores, how different is IRT from CTT? Has IRT been over-promoted as being the "modern" test theory over the "old" (classical) test theory when the Rasch model and CTT produce the same ordering of item difficulties and person abilities?

Using data from a large-scale statewide assessment, Fan (1998) found that person and item statistics obtained from CTT and IRT were quite comparable. In addition, the invariance property of item statistics across samples also appeared to be similar under the CTT and IRT frameworks. Further, Wainer (2007) made some interesting remarks in a book review of the fourth edition of Educational Measurement (Edited by Brennan 2006):

> I judge that at least 80% of all psychometric demands at the Educational Testing Service could be well handled with the material in Gulliksen's (1950/1978) classic text. Of the remaining 20% (primarily CAT work that is most gracefully done with IRT), most is covered in Lord and Novick (1968, p. 485).

In fact, classical test theory provides useful tools for analysing item responses. In particular, when we are dealing with a single test, item difficulty estimates and item discrimination index from CTT provide adequate information for test writers to select items for a test. The CTT discrimination index essentially provides comparable information as the residual-based IRT fit statistics (see Chap. 8). Test scores provide reasonable ability estimates particularly when tests are long (say, over 50 items).

However, IRT becomes useful when we are dealing with multiple tests where equating is needed. This could be for monitoring trends over time, or for estimating achievement differences between grade levels, or for computer adaptive testing, or for rotated test forms in one test administration. Further, IRT provides a more convenient way to link ability levels to item difficulties so that learning progressions can be more easily constructed. But CTT statistics should still form an essential part of the item analysis, and should be used to complement IRT analyses.

Fit of Data to the Rasch Model

The nice properties of the Rasch model discussed so far only hold if the item response data fit the model. That is, if the data do not fit the Rasch model, by applying a Rasch scaling, the property of specific objectivity will not hold. Therefore, to claim the benefit of using the Rasch model, the data must fit the model to begin with. Applying the Rasch model cannot "fix" problematic items! From this point of view, the use of the Rasch model for selecting items in the pilot stage of an assessment is most important. If the item response data from the final form of a test do not fit the Rasch model, the scale construction will not have sample invariance properties even when the Rasch model is applied. Chapter 8 discusses the evaluation of item fit to the Rasch model.

Estimation of Item Difficulty and Person Ability Parameters

Chapter 6 discusses the idea of estimating ability when the items are placed in their difficulty order (e.g., see Fig. 6.10). Essentially, when item difficulties are known, a student will likely answer around 50% of the items correctly for the set of items with difficulty at the ability level of the student. That is, given a student's item responses, to find a student's ability, we find the "mostly likely" ability value at which the observed item responses would occur. This is the notion of the maximum likelihood estimation method.

There are several different estimation methods to estimate item difficulty and ability parameters. In the following, we describe the idea of the joint maximum likelihood estimation (JML) method (e.g. Wright and Panchapakesan 1969; Linacre 1998). Other estimation methods such as conditional maximum likelihood (CML) and marginal maximum likelihood methods (MML) are also frequently used (e.g. Molenaar 1995; Baker and Kim 2004; de Ayala 2009).

First, consider the case when item difficulty estimates (δ_i) are known but abilities are unknown. To find the most likely ability for a student, we can form probability statements about the likelihood of observing the set of item responses. For example, if the ability is θ, we can compute the probability of obtaining a particular item response for various values of θ by using the formula

$$P(X = 1) = \frac{\exp(\theta - \delta_i)}{1 + \exp(\theta - \delta_i)} \quad \text{for correct response on Item } i \text{, and}$$

$$P(X = 0) = \frac{1}{1 + \exp(\theta - \delta_i)} \quad \text{for incorrect response on Item } i.$$

If we multiply together the individual probabilities for all item responses for the student on a test, we obtain the probability for the set of item responses. We then find the θ value that will maximise this probability. That is, we find the ability value that makes the observed item response pattern most likely. This process is repeated for each student to estimate abilities for the group of students.

Of course the problem with this approach is that initially we know neither the item difficulty values nor the abilities of students. So initially, we set some "reasonable" guesses to the item difficulty values, such as using formula $\log\left(\frac{p}{1-p}\right)$ where p is the percentage of students who obtained the correct answer on an item (see earlier discussion in this chapter). Using this set of item difficulty estimates, proceed to estimate abilities as described above. Using the new set of abilities, we can re-estimate item difficulties using an equivalent process as for ability estimates. That is, find item difficulty values to maximise the probability (likelihood) of observed data given the current ability estimates. Once new item difficulty estimates are obtained, re-estimate abilities using the updated item difficulties, so the iterations continue until the change in estimated parameter values between iterations is small, thereby indicating that convergence is achieved.

Weighted Likelihood Estimate of Ability (WLE)

The ability estimates obtained using the JML method as described above are often referred to as MLE (maximum likelihood estimate). MLE are found to be biased outwards. To remove this bias, Warm (1989) proposed a correction, and he called the corrected estimate Weighted Likelihood Estimate (WLE). Most IRT software programs provide WLE as ability estimates. WLE is also our preferred ability estimates for individual students. Linacre (2009) provides some comparisons between WLE and MLE.

The following is a note about zero and perfect scores. When students obtain zero or perfect scores, the maximum likelihood estimation has no solution, since if a student obtains no correct answer to any question, the ability of the student could be "infinitely" low, and there is no information about how low the ability is. Similarly, a student obtaining perfect score on a test could have a very high ability, and there is no information about how high the ability could go up. For these reasons, using

MLE for ability estimates, there often is an arbitrary convention of setting ability estimates for zero and perfect scores. For the WLE though, the correction made to MLE overcomes the estimation problem for zero and perfect scores, so that WLE ability estimates can be computed for zero and perfect test scores.

Local Independence

An important assumption is made in the specification of the Rasch probability function and the estimation procedure described in the previous section. That is, the probability of success $\frac{\exp(\theta_n-\delta_i)}{1+\exp(\theta_n-\delta_i)}$ depends (only) on a person's ability, θ_n, and an item's difficulty, δ_i. The probability is not influenced by a person's success or failure on other items, or by factors other than ability and item difficulty (such as particular person attribute and item characteristic). When this assumption holds, the likelihood of a set of observed item responses can be computed as the product of the probabilities of individual responses. This property is sometimes known as "local independence". The term "local" refers to the probability conditional on the values of θ_n and δ_i.

Violations of local independence can occur when there is "dependency" between items. For example, if item A cannot be answered unless the correct answer to item B is obtained, then the probability of success on item A is 0 if item B is incorrectly answered. In this case, Eq. (7.2) no longer holds. Testlets (a set of items linked to a common stimulus) often lead to the violations of local independence. It is likely that students' responses are more "similar than the model predict" on the items within a testlet, because of familiarity or otherwise with the testlet stimulus. So, in comparison with items outside the testlet, a student is more (or less) likely to obtain the correct answer than the probabilities predicted by the model.

Some fit tests are designed to check for the assumptions of local independence, as discussed in the next chapter.

Transformation of Logit Scores

While the raw score expressed in terms of percentage correct conveys the proportion of questions answered correctly, the logit scale in IRT does not have any "absolute" meaning in the sense that a logit of 1.5 is quite meaningless by itself. However, if a comparison is made, e.g., 1.5 logit is the difference between two students' abilities, then probability statements can be made about the relative chances of success of these two students on items. In general, the setting of the location of zero and the scale factor on a logit scale is quite arbitrary. For example, for convenience, we may set the average of all item difficulties at the zero point on the logit scale, or we may set students' average ability as the zero point on the logit scale. For a scale factor, we may constrain the variance of the student abilities to 1 or set the standard deviation to some arbitrary number.

Since logit values do not have any absolute meaning by themselves, and since logit values can be zero and negative, logits have often been transformed to avoid mis-interpretations, particularly for zero or negative logits. For example, zero logit does not mean there is no ability. A negative logit value does not mean worse than no ability. A logit value of 100 does not mean maximum ability such as 100% correct. For these reasons, researchers often choose a linear transformation so that the range of student abilities is typically positive and above 100 so they are less likely to be confused with percentages of items correct. As an example, in PISA, a linear transformation of IRT abilities is used so that the mean of student abilities is set at 500 and the standard deviation is set at 100. In this way, the typical range of student abilities is between 300 and 700. These values are positive so we don't need to explain about negative ability. They are away from zero so there is no mis-interpretation of zero logit. They are away from 0 to 100 so they won't be confused with percentages. As the transformation is linear, the transformation does not impact on the interval property of the scores.

An Illustrative Example of a Rasch Analysis

The data set for this example comes from a numeracy practice test for Grade 3 students. There are 15 test items and 1199 respondents. The analysis is carried out with R package TAM (Kiefer et al. 2013). For installing R and the TAM package, see website

www.edmeasurementsurveys.com/TAM/Tutorials.

The R scripts for this analysis are shown in Table 7.1.

Line 1 of the R script file loads the TAM package for IRT analysis.
Line 2 sets the working directory. The working directory is the default folder for all files if no full directory path is specified.
Line 3 reads the data file in csv format. The data file is in the working directory.
Line 4 calls the TAM function "tam.jml2" to run a JML analysis. The results of the run are encapsulated in an R object called "mod1". All results are accessed via "mod1" using syntax "mod1$xyz" where "xyz" is a specific variable name of the IRT output. See TAM pdf documentation of the output variable names from JML.
Line 5 displays the item difficulty parameters (δ in Eq. (7.1)), shown in Table 7.2. Table 7.2 shows that Item 7 is the most difficult item, while Item 1 is the easiest.
Line 6 of the R script displays the WLE ability estimates. An excerpt is shown in Table 7.3.
As the data file has been sorted in test score order, the first 15 respondents have test scores of 0, 1 and 2. It can be seen that students with the same test score have the same WLE ability estimates, illustrating the idea of raw score as "sufficient statistic" for ability estimate.

Table 7.1 R code for an IRT analysis using R package TAM

	R script	Comment
1	`library("TAM")`	Load R library "TAM"
2	`setwd("C:/G_MWU/IRTBook/Chapter7_RaschModel/data")`	Set working directory. Note forward slash "/", not backward slash "\"
3	`resp <- read.csv("NumeracyA1.csv")`	Read data file in csv format
4	`mod1 <- tam.jml2(resp)`	Run IRT analysis using JML
5	`mod1$xsi`	Show IRT item difficulty estimates
6	`mod1$WLE`	Show IRT ability estimates
7	`plot(mod1)`	Plot ICC
8	`ctt <- tam.ctt(resp, mod1$WLE)`	Classical test theory statistics
9	`ctt`	Show CTT statistics
10	`score <- rowSums(resp)`	Compute raw scores
11	`plot(score, mod1$WLE)`	Plot raw score against WLE

Table 7.2 Item difficulty parameter estimates

Item difficulty parameters	
1	−3.597902770
2	0.152710768
3	−0.216339997
4	−0.315478176
5	−0.004676182
6	0.269804125
7	2.491833957
8	−2.005791842
9	−0.720173111
10	−0.424414757
11	−0.866854422
12	−1.244359936
13	−1.938983421
14	−0.336291154
15	1.117124898

Line 7 of the R script file plots the item characteristic curves (ICC). Four ICCs are shown as examples in Fig. 7.7.

Each item in Fig. 7.7 shows two ICCs: the smooth curve (blue) is the expected scores curve while the joined line segments (five segments joining 6 points) show the observed ICC from empirical data. Item 7 (top right graph) shows a difficult

Table 7.3 Excerpt of WLE ability estimates

Excerpt of WLE ability	
1	−4.608557
2	−4.608557
3	−4.608557
4	−4.608557
5	−4.608557
6	−3.700321
7	−3.700321
8	−3.700321
9	−3.700321
10	−3.700321
11	−3.700321
12	−3.700321
13	−3.700321
14	−2.864732
15	−2.864732

item, with ICCs low in the graph, indicating low probabilities of being successful on this item for all ability levels. Items 4, 9 and 10 are medium level difficulty items. However, Item 4 shows that the observed ICC matches the expected ICC well. Item 9 shows the observed ICC flatter than the expected ICC. Item 10 shows the observed ICC steeper than the expected ICC.

For Item 9 (bottom left graph), low ability students have a higher chance of obtaining the correct answer than expected, and fewer high ability students obtained the correct answer than expected, leading to a somewhat flatter observed ICC. In general, a very flat ICC indicates that the item does not discriminate between high and low ability students well, since the probability of obtaining a correct answer does not differ greatly for low and high ability students.

In contrast, Item 10 discriminates low- and high-ability students better than "what the model expects" in that low ability students are quite likely to answer the item incorrectly, while high ability students are mostly answering the item correctly. In short, Item 9 has low discriminating power as compared to the other items in the test, and Item 10 has high discriminating power as compared to the other items. Figure 7.8 shows the actual items (Item 9 (top) and Item 10 (bottom)).

Item 9 is a question about the use of terms "left-right" and "top-bottom". Item 10 is a computation problem where division is involved. One can make conjectures about why Item 9 has low discrimination and why Item 10 has high discrimination. Looking through the whole test, Item 11 is also a division problem similar to Item 10 and it also has high discrimination. One may conjecture that computation items involving division discriminate better between low and high achievers, while using and understanding of terms such as "left-right-top-bottom" may not be the best indicator of students' mathematics abilities. Such content analysis forms part of the item analysis, and subject experts need to be called on to make these interpretations.

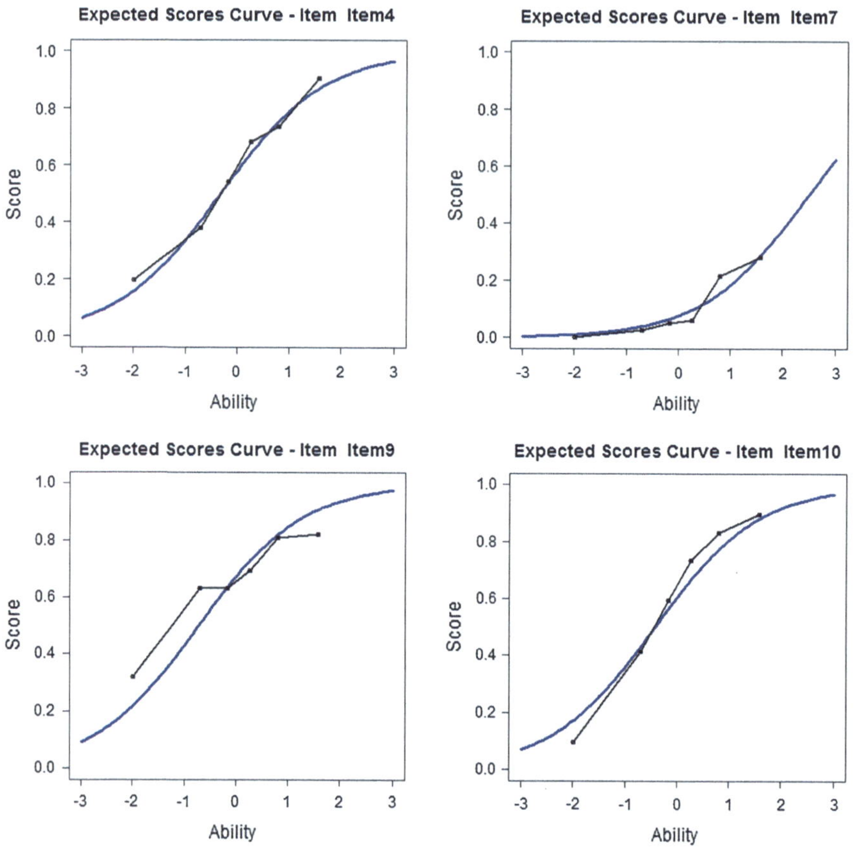

Fig. 7.7 ICCs for four items

Statistics alone can only inform us of numerical values of variables of interest but not the implications of these numbers.

To further check the discriminating powers of the items in the test, a classical test theory function in the TAM package is run. Line 8 of the R script file calls the CTT function and line 9 prints the results. A summary of the CTT analysis for Items 9, 10 and 11 is shown in Table 7.4.

In Table 7.4, column "AbsFreq" shows the number of students in each item response category, and column "RelFreq" shows the frequency in percentage. Column "rpb.WLE" shows the CTT discrimination index. It can be seen that Item 9 has a rpb (point-biserial correlation) of 0.36, while Item 10 has a rpb of 0.54 and Item 11 has a rpb of 0.55. The column headed "M.WLE" shows the average WLE ability of students (from IRT analysis) in each item response category. The average ability of students obtaining the correct answer for Item 9 is 0.28, while the average abilities for Item 10 and Item 11 are 0.50 and 0.42 respectively. So the CTT analysis corroborates the results of the IRT analysis.

Fig. 7.8 Item 9 (*top*) and Item 10 (*bottom*) in the example test

Table 7.4 CTT analysis for Items 9, 10 and 11

Item	N	Categ	AbsFreq	RelFreq	rpb.WLE	M.WLE
Item 9	1199	0	418	0.348624	−0.35937	−0.62141
Item 9	1199	1	781	0.651376	0.359374	0.282965
Item 10	1199	0	486	0.405338	−0.54255	−0.82038
Item 10	1199	1	713	0.594662	0.542548	0.504841
Item 11	1199	0	386	0.321935	−0.54992	−0.9894
Item 11	1199	1	813	0.678065	0.549923	0.422085

Lines 10 and 11 of the R script demonstrate the relationships between raw test score and IRT ability estimates. Line 10 computes test scores for students by adding the scores across each row of the response matrix, as each row contains the item scores for each student. Line 11 plots the test scores against the IRT scores. The graph is shown in Fig. 7.9.

Figure 7.9 shows that there is a one-to-one relationship between test score and IRT ability estimate. The relationship is not a straight line (i.e., not linear); it is the

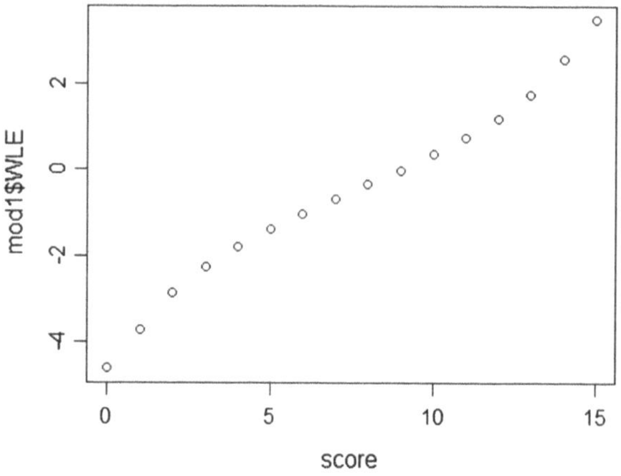

Fig. 7.9 Plot of test raw score versus IRT ability estimate

shape of a log function. The IRT ability estimates spread out a little more at the lower and upper ends of the test scores range. Nevertheless, for test scores in the middle range (say, between 5 and 10), the relationship is close to a straight line.

Summary

This chapter formally introduces the Rasch model for the dichotomous case. Various properties of the Rasch model are discussed including specific objectivity, equal item discrimination, difficulty ordering, raw scores as sufficient statistics, and JML estimation method. The main message is that the Rasch model provides very desirable measurement properties. The notion of item difficulty under the Rasch model is clearly defined in the sense that items can be ordered by their item difficulty parameters and such an ordering is valid for students at all ability levels. This is not the case for 2- and 3-parameter IRT models. Similarly, the relative differences between the abilities of students can be made without references to specific items.

However, while in theory the Rasch model has many good measurement properties, in practice, classical test theory and other IRT models all have their uses and provide complementary information. For a broader overview and comparisons of IRT models, see Thissen and Steinberg (2009).

Hands-on Practices

Task 1

In EXCEL, compute the probability of success under the Rasch model, given an ability measure and an item difficulty measure. Plot the item characteristic curve. Follow the steps below.

Step 1 In EXCEL, create a spreadsheet with the first column showing abilities from −3 to 3, in steps of 0.1. In Cell B2, type in a value for an item difficulty, say 0.8, as shown below.

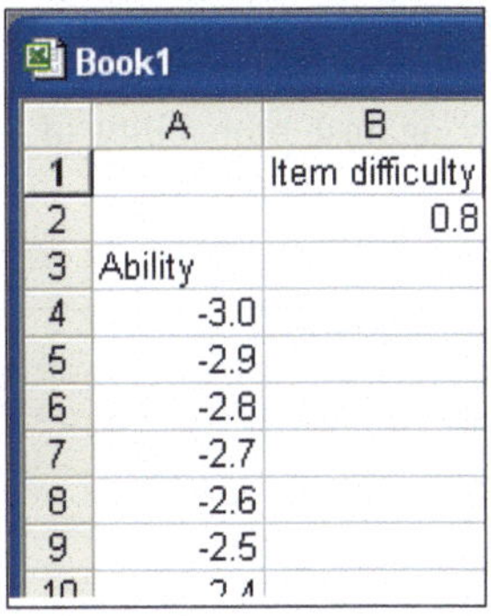
Book1

	A	B
1		Item difficulty
2		0.8
3	Ability	
4	-3.0	
5	-2.9	
6	-2.8	
7	-2.7	
8	-2.6	
9	-2.5	

Step 2 In Cell B4, compute the probability of success: Type the following formula, as shown

$$= \exp(\$A4 - B\$2)/(1 + \exp(\$A4 - B\$2))$$

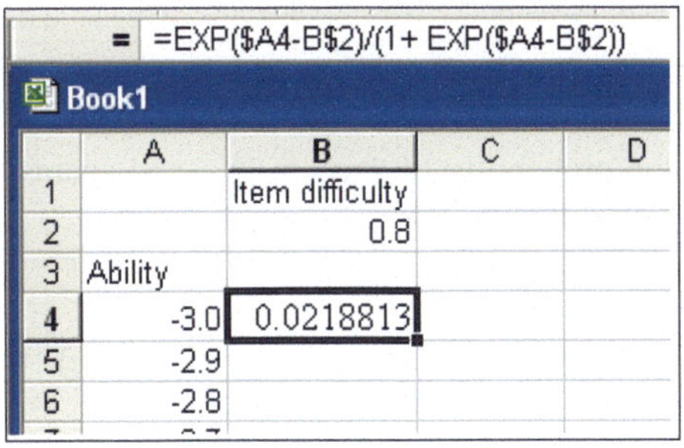
= =EXP($A4-B$2)/(1+ EXP($A4-B$2))

Book1

	A	B	C	D
1		Item difficulty		
2		0.8		
3	Ability			
4	-3.0	0.0218813		
5	-2.9			
6	-2.8			

Step 3 Autofill the rest of column B, for all ability values, as shown

Book1

	A	B
1		Item difficulty
2		0.8
3	Ability	
4	-3.0	0.0218813
5	-2.9	0.024127
6	-2.8	0.026597
7	-2.7	0.0293122
8	-2.6	0.0322955

Step 4 Make a XY (scatter) plot of ability against probability of success, as shown below.

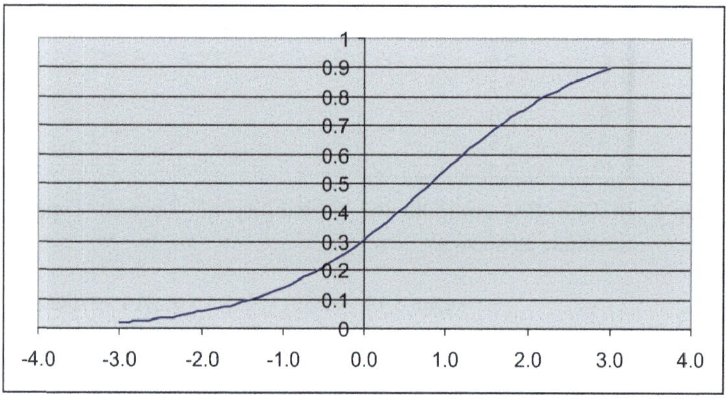

This graph shows the probability of success (Y axis) against ability (X axis), for an item with difficulty 0.8.

Q1. When the ability equals the item difficulty (0.8 in this case), what is the probability of success?

Step 5 Add another item in the spreadsheet with item difficulty of −0.3. In Cell C2, enter −0.3. Autofill cell C4 from cell B4. Then autofill the column of C for the other ability values.

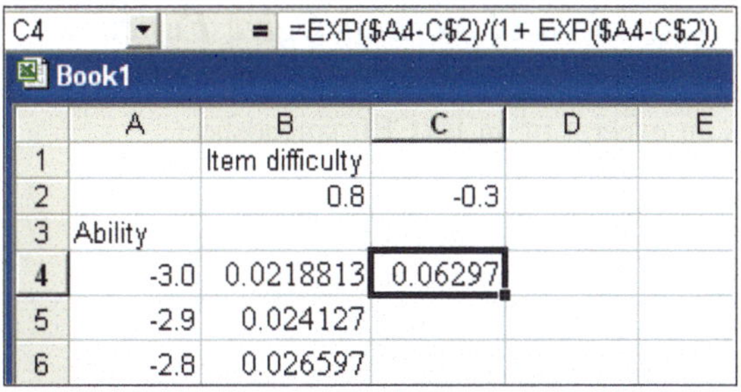

	A	B	C	D	E
1		Item difficulty			
2		0.8	-0.3		
3	Ability				
4	-3.0	0.0218813	0.06297		
5	-2.9	0.024127			
6	-2.8	0.026597			

Step 6 Plot the probability of success on both items, as a function of ability (hint: plot columns A, B and C).

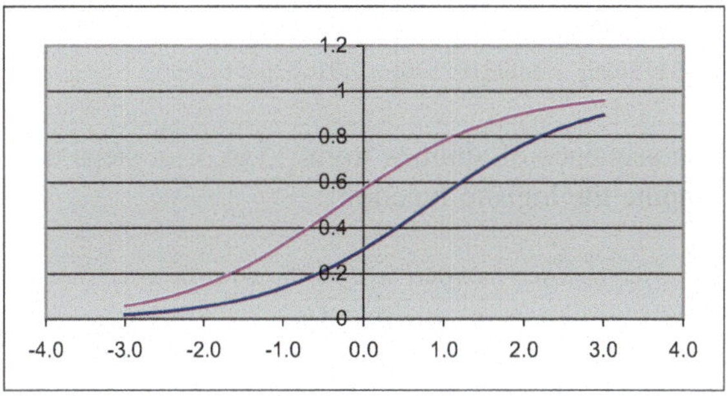

Q2. A person with ability −1.0 has a probability of 0.14185 of getting the first item right. At what ability does a person have the same probability of getting the second item right?

Q3. What is the difference between the abilities of the two persons where Person A's probability of getting the first item right is the same as Person B's probability of getting the second item right?

Q4. How does this difference relate to the item difficulties of the two items?

Q5. If there is a very difficult item (say, with difficulty value of 2), can you sketch the probability curves on the above graph (without computing it in EXCEL)? Check your graph with an actual computation and plot in EXCEL.

Task 2. Compare Logistic and Normal Ogive Functions

The following shows an excerpt of an EXCEL spreadsheet.

C2 fx =NORMDIST(A2,0,1,1)

	A	B	C	D
1	ability	logistic	normal	logistic1.7
2	-3	0.047425873	0.001349898	0.006059801
3	-2.9	0.052153563	0.001865813	0.007174656
4	-2.8	0.057324176	0.00255513	0.008492863
5	-2.7	0.062973356	0.003466974	0.010050814
6	-2.6	0.06913842	0.004661188	0.011891132
7	-2.5	0.07585818	0.006209665	0.014063627
8	-2.4	0.083172696	0.008197536	0.016626356

In column A, set a sequence of abilities from −3 to 3, in steps of 0.1.

In column B, compute the logistic function

$$\frac{\exp(\theta - \delta)}{1 + \exp(\theta - \delta)}$$

where θ is the person-parameter (ability in column A) and δ is zero. More specifically, type the formula "=exp(A2)/(1 + exp(A2))" in cell B2, and auto-fill column B.

In column C, compute the normal ogive function (i.e., cumulative normal distribution with mean 0 and standard deviation 1). Specifically, in column C2, type the formula "=normdist(A2, 0, 1, 1)".

In column D, compute the logistic function, but this time, use a scale parameter of 1.7. Specifically, in cell D2, type the formula "=exp(1.7 * A2)/(1 + exp(1.7 * A2))".

Make a scatter plot of columns A to D on the same graph. You should get a graph like the following:

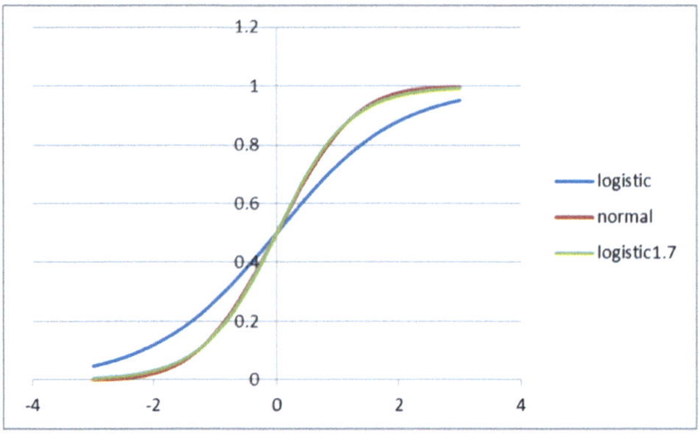

Which two functions overlap with each other?

Since the ability scale has an arbitrary scale factor, it does not matter whether we use $\frac{\exp(\theta-\delta)}{1+\exp(\theta-\delta)}$ or $\frac{\exp(1.7(\theta-\delta))}{1+\exp(1.7(\theta-\delta))}$.

The latter form is still a Rasch model, and it is very close to the normal ogive function. For this reason, some software programs set the ability scale with a scale factor of 1.7 instead of 1. This will not affect the interpretations of IRT results in any way.

Task 3. Compute the Likelihood Function

To understand the idea of raw score as sufficient statistic for ability estimate, this hands-on practice shows the comparison between likelihood functions for different item response patterns with the same raw test score.

We will use EXCEL for this exercise. Figure 7.10 shows an EXCEL spreadsheet.

In this example, a test has five items where the item difficulties are −2, −1, 0, 1 and 2 (see Row 2 in the spreadsheet in Fig. 7.10). In cells A5 to A17, there is a list of abilities from −3 to 3. Row 3 contains an item response pattern. In this example, it is assumed that a student obtained a score of three by answering Items 1, 3 and 5 correctly. In cells B5 to F17, compute the probability of the item response with given ability. The EXCEL

B5 fx =EXP(B$3*($A5-B$2))/(1+EXP($A5-B$2))

	A		B	C	D	E	F	G
1			Item1	Item2	Item3	Item4	Item5	likelihood function
2	item difficulty		-2	-1	0	1	2	
3	item response		1	0	1	0	1	
4	ability							
5		-3	0.268941	0.880797	0.047426	0.982014	0.006693	7.38376E-05
6		-2.5	0.377541	0.817574	0.075858	0.970688	0.010987	0.000249718
7		-2	0.5	0.731059	0.119203	0.952574	0.017986	0.00074653
8		-1.5	0.622459	0.622459	0.182426	0.924142	0.029312	0.001914675
9		-1	0.731059	0.5	0.268941	0.880797	0.047426	0.004106493
10		-0.5	0.817574	0.377541	0.377541	0.817574	0.075858	0.007227441
11		0	0.880797	0.268941	0.5	0.731059	0.119203	0.010321496
12		0.5	0.924142	0.182426	0.622459	0.622459	0.182426	0.011916036
13		1	0.952574	0.119203	0.731059	0.5	0.268941	0.011162605
14		1.5	0.970688	0.075858	0.817574	0.377541	0.377541	0.008580978
15		2	0.982014	0.047426	0.880797	0.268941	0.5	0.005516155
16		2.5	0.989013	0.029312	0.924142	0.182426	0.622459	0.003042188
17		3	0.993307	0.017986	0.952574	0.119203	0.731059	0.001483068

Fig. 7.10 Computation of likelihood function

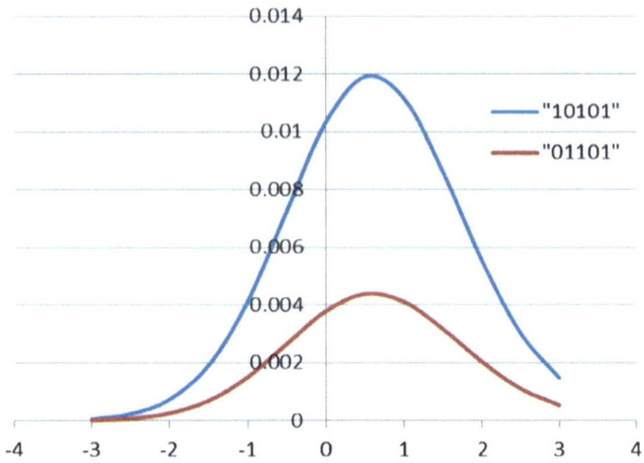

Fig. 7.11 Likelihood as a function of ability and item response pattern

formula in B5 is "=EXP(B$3 * ($A5 − B$2))/(1 + EXP($A5 − B$2))". The multiplier "B$3" in the formula is the item response. This formula evaluates to $\frac{\exp(\theta-\delta_i)}{1+\exp(\theta-\delta_i)}$ when the item response is 1, and $\frac{1}{1+\exp(\theta-\delta_i)}$ when the item response is 0. Column G is the likelihood which is the product of the probabilities in cells B to F.

Scanning down column G, the largest value (maximum) is in row 12, when the ability is 0.5. This is the ability that "maximises" the likelihood. So the ability estimate for a response pattern of "10101" (right-wrong-right-wrong-right) will be around 0.5.

Repeat this computation for a different response pattern but still for a raw score of 3. For example, use the response pattern "01101" in cells B3 to F3. What is the maximum likelihood and which ability corresponds to this maximum likelihood?

Plot the likelihood functions for both response patterns "10101" and "01101" as a function of ability. Figure 7.11 shows such a graph.

Repeat this for another response pattern, say, "11100", and plot the likelihood for the three response patterns on the same graph.

Discussion Points

1. In relation to the Hands-on Practices Task 3, discuss the concept of "raw score as sufficient statistic for ability estimate". For example, given a raw score, how do different response patterns affect the estimation of ability? Given a raw score, compare the likelihood functions for similarities and differences for different response patterns. Would different response patterns impact on the fit of the items to the model?
2. From a fairness point of view, do you think the ability estimate should be the same for the same raw score irrespective of the actual items answered correctly? If not, how would you score the items? Consider the illustrative example. If Items 9 and 10 are administered, one student got Item 9 right but Item 10 wrong,

the other student got Item 9 wrong but Item 10 right, so both students got 1 out of 2. Should they have the same ability estimate?

3. Discuss the concept of "sample-free" in Rasch models. In what ways are estimated statistics sample-free and sample-dependent?

References

Baker FB, Kim S-H (2004) Item response theory: parameter estimation techniques, 2nd edn. Dekker, New York

Birnbaum A (1968) Some latent trait models and their use in inferring an examinee's ability. In: Lord FM, Novick MR (eds) Statistical theories of mental test scores. Addison-Wesley, Reading, pp 395–479

de Ayala RJ (2009) The theory and practice of item response theory. The Guilford Press, New York

Embretson SE, Reise SP (2000) Item response theory for psychologists. Lawrence Erlbaum Associates, Mahwah

Fan X (1998) Item response theory and classical test theory. An empirical comparison of their item/person statistics. Educ Psychol Measur 58(3):357–381

Humphry SM, Andrich D (2008) Understanding the unit in the Rasch model. J Appl Measur 9 (3):249–264

Kiefer T, Robitzsch A, Wu M (2013) TAM (Test analysis modules)—an R package [computer software]

Linacre JM (1998) Estimating Rasch measures with known polytomous item difficulties. Rasch Measur Trans 12(2):638

Linacre JM (2009) The efficacy of Warm's weighted mean likelihood estimate (WLE) correction to maximum likelihood estimate (MLE) bias. Rasch Measur Trans 23(1):1188–1189

Lord FM, Novick MR (1968) Statistical theories of mental test scores. Addison-Wesley, Reading

Molenaar IW (1995) Estimation of item parameters. In: Fischer G, Molenaar IW (eds) Rasch models: foundations, recent developments, and applications. Springer, New York, pp 39–51

Rasch G (1960) Probabilistic models for some intelligence and attainment tests. Danish Institute for Educational Research, Copenhagen

Rasch G (1977) On specific objectivity: an attempt at formalizing the request for generality and validity of scientific statements. The Danish Yearbook of Philosophy 14:58–93

Samejima F (1977) The use of the information function in tailored testing. Appl Psychol Meas 1:233–247

Thissen D, Steinberg L (2009) Item response theory. In: Millsap RE, Maydeu-Olivares A (eds) The Sage handbook of quantitative methods in psychology. Sage, Thousand Oaks, pp 148–177

Thissen D, Wainer H (2001) Test scoring. Lawrence Erlbaum Associates, Mahwah

van der Linden WJ, Hambleton RK (1997) Handbook of modern item response theory. Springer, New York

Verhelst ND, Glas CAW (1995) One-parameter logistic model. In: Fischer G, Molenaar IW (eds) Rasch models: Foundations, recent developments, and applications. Springer, New York, pp 215–237

Wainer H (2007) A psychometric cicada: educational measurement returns. Book review. Educ Researcher 36:485–486

Warm TA (1989) Weighted likelihood estimation of ability in item response theory. Psychometrika 54:427–450

Wilson M (2011) Some notes on the term: "Wright Map". Rasch Measur Trans 25(3):1331
Wright BD (1977) Solving measurement problems with the Rasch model. J Educ Meas 14:97–115
Wright BD, Panchapakesan N (1969) A procedure for sample-free item analysis. Educ Psychol Measur 29:23–48

Further Readings

Baker F (2001) The basics of item response theory. ERIC Clearinghouse on Assessment and Evaluation, University of Maryland, College Park, MD. Available online at http://edres.org/irt/baker/
Fischer GH, Molenaar IW (eds) (1995) Rasch models: foundations, recent developments, and applications. Springer, New York
Hambleton RK, Swaminathan H, Rogers HJ (1991) Fundamentals of item response theory. Sage, Newbury Park
Harris D (1989) Comparison of 1-, 2-, and 3-parameter IRT models. NCME Instructional Topics in Educational Measurement Series (ITEMS) Module 7. Retrieved 21 July 2014 from http://ncme.org/publications/items/. There are other modules at this website on various topics of educational measurement
Rasch G (1980) Probabilistic models for some intelligence and attainment tests. University of Chicago Press, Chicago
Rasch Measurement Transactions. http://www.rasch.org/rmt. Many helpful articles can be found at this website

Chapter 8
Residual-Based Fit Statistics

Introduction

This chapter discusses the commonly used residual-based fit statistics for the Rasch model. These item- and person-fit statistics are derived to check how well the item response data fit the Rasch model. We note that these fit statistics do not assess the overall model fit. Instead, they provide fit indices for each item and for each person. There has been a great deal of confusion in using these statistics so it is worth devoting a chapter to describe their appropriate uses. It should be stressed that there are many different fit indices, some of which check for overall model fit (e.g. see Maydeu-Olivares (2013) for an overview of different methods), and some fit indices check for specific violations of the model. For example, there are fit statistics for checking the violation of the unidimensional assumption (e.g. McDonald and Mok 1995; Hattie 1985). There are also fit statistics for checking the violation of the assumption of local independence between items (e.g. Liu and Maydeu-Olivares 2013; Chen and Thissen 1997). The residual-based fit statistics presented here are just one form of fit indices for items and for persons, derived using differences between the observed item score and the expected item score to form residuals for each item and for each person.

While the Rasch model described in Chap. 7 has many good measurement properties, there is no guarantee that the item response data collected will conform to the mathematical formulation of the Rasch model. If the data collected do not conform to the Rasch model, the application of the Rasch model will not improve the measurement properties of the data. That is, unless the data actually fit the Rasch model, there is little point in using the Rasch model. Therefore, it is important to assess the extent to which the data fit the Rasch model.

The key feature of the Rasch model is that the probability of success on an item can be completely determined by two values: an item difficulty, δ, and a person ability, θ. Equation (8.1) shows the Rasch model for the probability of success for a person on an item.

M. Wu et al., *Educational Measurement for Applied Researchers*,
DOI 10.1007/978-981-10-3302-5_8

$$p = P(X = 1) = \frac{\exp(\theta - \delta)}{1 + \exp(\theta - \delta)} \tag{8.1}$$

If there are factors other than the item difficulty and person ability that influence the probability of success for a person on an item, then the assumptions of the Rasch model are violated. Some of these factors may include the following:

- Guessing. Guessing can occur, particularly for difficult multiple choice items. In general, we often find that short-constructed items are more "discriminating" than multiple-choice items.
- Item Dependency. The "local independence" assumption of the Rasch model is violated when the probability of success on an item depends on the response(s) to other item(s). For example, an item requires information from the answer of a previous item, or, one item provides clues to the answer of another item.
- Differential Item Functioning (DIF). DIF occurs when different groups of students respond to an item in different ways. For example, boys may perform better than girls on an item about football because boys are more engaged with the sport.
- Other Traits. An item may tap into a number of "traits". For example, a mathematics item may be testing both conceptual understanding and computational accuracy. These two "traits" may be different for different individuals. That is, a person may be high on one trait, but low on the other. This is a form of a violation of the unidimensionality assumption in the IRT model.

Fit Statistics

The extent to which the Rasch model assumptions are violated can be tested through "fit statistics". However, since there are many factors that can affect the assumptions of the Rasch model, different fit statistics have been designed to detect different kinds of violations. This is an important point to remember, as too often we make judgements based on a single fit statistic about whether data fit the Rasch model. It should be noted that each fit statistic is sensitive only to specific violations of the model, and may not be sensitive to other violations of the model.

Many fit indices have been developed for testing the adequacy of item response models and for exploring the properties of test items. Douglas (1982) discussed the notion of *fit*, from both statistical and psychometric perspectives. He also presented a technical review of many of the fit statistics that have been developed, and compared a number of fit statistics, including Wright and Panchapakesan's (1969) between fit statistic, van den Wollenburg's (1982) Q1 statistic and Andersen's (1973) likelihood ratio test. Douglas demonstrated that a number of alternative statistics are in fact indistinguishable in practice, and concluded the review with the following observation:

> The most valuable contribution to the area of tests of fit for Rasch models in recent years has been the recognition by some psychometricians that there is no such thing as a final 'fit' of data to the model and hence that no one test is ever likely to be complete (p. 43).

In the years since Douglas' review, a considerable amount of new development work on fit statistics has been undertaken (for example, Glas and Verhelst 1995; Smith 1988). The increased power of computers has enabled more complex computations to be performed and simulation studies have been used to test certain conjectures and hypotheses. Where analytic derivations have become difficult, empirically based approaches have been applied to provide better insight into the properties of some of the fit statistics for which only the theoretical asymptotic properties are known.

Generally, there are three types of fit statistics:

(1) Chi-square goodness-of-fit tests that are based on comparing observed and expected counts of various types (e.g., Glas and Verhelst 1995).
(2) Tests that combine standardised residuals to form approximate normal variates. These tests are based on comparing the observed and expected responses of individuals to items (Wright and Panchapakesan 1969, Wright 1977).
(3) Exploratory and nonparametric tests that provide diagnostic information about specific model violations (e.g., Molenaar 1983, DIMTEST (Stout et al. 1996)).

Meijer and Sijtsma (2001) provided a comprehensive overview of various kinds of person fit statistics.

Residual-Based Fit Statistics

In this chapter, we will focus on one type of fit statistics: the residual-based fit statistics. This type of fit statistics is reported in a number of IRT software packages such as Winsteps (Linacre and Wright 2000), RUMM (2001), ConQuest (Wu et al. 1998), TAM (Kiefer et al. 2013). A detailed discussion of residual-based fit statistics can be found in Wu and Adams (2013).

Wright (1977) proposed several item fit and person fit statistics based on standardised residuals for the Rasch model. Let X_{ni} be the observed score for person n on item i, and P_{ni} be the probability of obtaining a correct response for person n on item i. X_{ni} is the random variable for the (scored) item response for person n on item i. We use capital letters to denote random variables and small letters to denote observed values of corresponding random variables. Then the standardised residual is defined as

$$z_{ni} = \frac{(x_{ni} - E(X_{ni}))}{(Var(X_{ni}))^{\frac{1}{2}}} \tag{8.2}$$

where $E(X_{ni})$ is the expected value of the item response, and $Var(X_{ni})$ is the variance of the item response. In the case of the dichotomous Rasch model, $E(X_{ni}) = P_{ni}$ and $Var(X_{ni}) = P_{ni}(1 - P_{ni})$. The standardised residuals have served as general diagnostic tools in the assessment of item and person fit. They are mostly presented as graphical displays to draw attention to problematic items/persons, rather than used as vigorous statistical tests for the fit of the model.

Squaring z_{ni} and summing over n (persons), a statistic is derived that can be used as a fit index for item i. Squaring z_{ni} and summing over i (items), a statistic is derived that can be used as a fit index for person n (Wright and Masters, 1982). For item fit, Wright and Masters proposed an unweighted and a weighted statistic (sometimes called the outfit and infit, or the unweighted total fit and weighted total fit). The unweighted fit mean-square (outfit) statistic is defined as

$$\text{Unweighted mean square (outfit)} = \frac{\sum_n z_{ni}^2}{N} = \frac{1}{N}\sum_n \frac{(x_{ni} - E(X_{ni}))^2}{Var(X_{ni})} \tag{8.3}$$

where N is the total number of respondents. The weighted fit mean-square (infit) statistic is defined as

$$\text{Weighted mean square (infit)} = \frac{\sum_n z_{ni}^2 Var(X_{ni})}{\sum_n Var(X_{ni})} = \frac{\sum_n (x_{ni} - E(X_{ni}))^2}{\sum_n Var(X_{ni})} \tag{8.4}$$

That is, the standardised residuals, z_{ni}, are weighted by $Var(X_{ni})$, and the denominator in Eq. (8.4) is the sum of the weights.

When certain assumptions are made, it can be shown that both the unweighted mean-square (outfit) and the weighted mean-square (infit) have expectations of one. The variances of the mean-square can also be computed. Wright and Panchapakesan (1969) indicated that both the weighted and the unweighted mean-square can be treated as chi-square variates. They also suggested the use of a cube root transformation (the Wilson-Hilferty transformation) of the mean-square to obtain a t statistic that has an approximate normal distribution so that a frame of reference can be established in testing the fit of the model.

Additional Notes

The term "weighted mean-square" is used to indicate that the square of the standardised residuals are weighted by the variance of the item response (See Eq. (8.4)). Each z_{ni}^2 is multiplied by $Var(X_{ni})$ in the numerator of Eq. (8.4). The denominator is the sum of the weights. In contrast, for unweighted mean-square (Eq. (8.3)), each z_{ni}^2 can be considered to have a weight of one (equal weight), and the denominator, N, is the sum of the weights.

There is a common sense justification for the weight, $Var(X_{ni})$, used in weighted mean-square. Essentially, when the item difficulty of an item is close to the ability of a person, $Var(X_{ni})$ is relatively large. When an item is "off-target" (too easy or too hard), $Var(X_{ni})$ is relatively small. So one uses a larger weight when an item provides more "information" about an item or student (an on-target item), and one uses a smaller weight when an item does not provide much "information" about the item or person (an off-target item).

Example Fit Statistics

Figure 8.1 shows an example output showing values of fit mean-square and t statistics for each item. It can be seen that both the outfit and infit mean-square values are centred around one, and the t values are centred around zero.

Interpretations of Fit Mean-Square

While it is stated that the fit mean-square value has an expectation of one, we need to make an assessment of how far away the mean-square value is from one before we conclude that an item is regarded as a misfitting item. Further, when an item shows misfit, we need to understand the meaning of "over-fit" (mean-square value less than one) and "under-fit" (mean-square value greater than one).

Equal Slope Parameter

The mean-square statistic defined in Eq. (8.3) tests whether item i has the same "slope" as the other items in the test, since the Rasch model makes the assumption that all items have the same slope, or the same "discrimination" parameter value.

	outfitItem	outfitItem_t	infitItem	infitItem_t
Item1	1.0961083	0.41749661	0.9825207	-0.10583654
Item2	0.9995883	0.01182956	0.9972270	-0.10039567
Item3	0.9759595	-0.43143661	0.9943946	-0.19965112
Item4	0.9627763	-0.67759598	0.9990423	-0.02541451
Item5	0.9269022	-1.34523240	0.9585211	-1.62002975
Item6	0.8847570	-1.99894973	0.9421925	-2.30266170
Item7	0.7691715	-1.19101948	0.9492248	-0.71737778
Item8	1.0996387	0.84683135	0.9980673	-0.01535037
Item9	1.2471000	3.83770123	1.1645839	4.89218718
Item10	0.8914894	-2.02992537	0.9012658	-3.55879474
Item11	0.8428774	-2.57972225	0.8900627	-3.34304054
Item12	0.9568208	-0.53966836	1.0663672	1.62808234
Item13	0.9061466	-0.80405626	0.9519401	-0.87181921
Item14	1.0823112	1.49413443	1.0767440	2.69877980
Item15	1.2107439	2.18638272	1.0495056	1.50255396

Fig. 8.1 Example output from TAM software showing residual-based fit indices

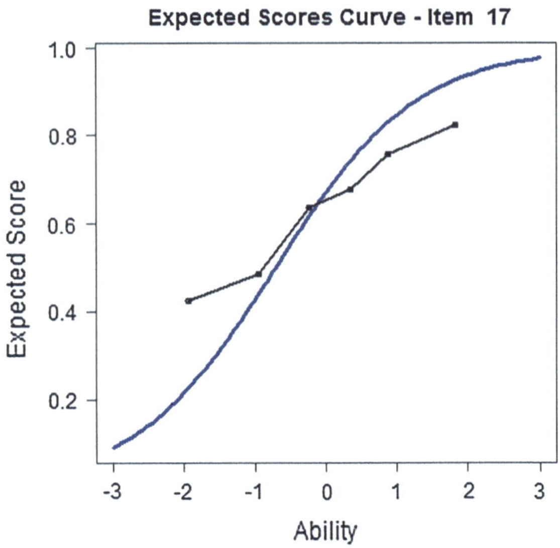

Fig. 8.2 Observed ICC is "flatter" than expected ICC (under-fit) (infit MNSQ = 1.27)

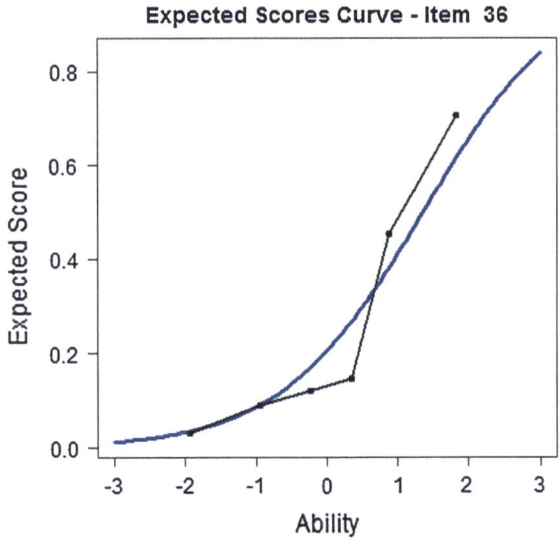

Fig. 8.3 Observed ICC is "steeper" than expected ICC (over-fit) (infit MNSQ = 0.90)

It can be shown that, when the observed item characteristic curve (ICC) is "steeper" than the expected ICC, the fit mean-square value is less than one. When the observed ICC is "flatter" than the expected ICC, the fit mean-square value is greater than one. Figure 8.2 shows an example where the observed ICC is flatter than the expected ICC (infit mean-square = 1.27). Figure 8.3 shows an example where the observed ICC is steeper than the expected ICC (infit mean-square = 0.90). (See Wu and Adams (2013) for more detailed mathematical explanations).

We note that the slope of the expected (or theoretical) ICC is the "average" of the slopes of all observed ICCs. So, in every data set, if some items show "under-fit", some items will show "over-fit".

Not About the Amount of "Noise" Around the Item Characteristic Curve

Contrary to common belief, the residual-based fit statistics do not provide an indication of how far away the observed ICC is from the theoretical ICC. That is, provided that the "slope" of the observed ICC is the same as the slope of the theoretical ICC, the fit mean-square will not show misfit whether the observed ICC is close or far away from the theoretical ICC.

Figure 8.4 shows an item where the observed ICC appears to be close to the theoretical ICC for all ability groups. The weighted fit mean-square is 1.01. By contrast, Fig. 8.5 shows an item where the observed ICC has a number of points "far away" from the theoretical ICC, particularly for ability groups in the higher range. Yet the weighted fit mean-square is still 1.00. These two examples show that the fit mean-square statistic is not about the amount of "noise" of the observed ICC as compared to the theoretical ICC. Rather, the fit mean-square statistic is testing whether the "slope" of the observed ICC is the same as the theoretical ICC.

It is worth stressing the point that the Rasch model does not specify an absolute value for the discrimination parameter. Therefore, when an item is identified as a misfitting item, it shows that the item is different from the other items in the same test. So from this point of view, the "fit" index shows "relative fit" and "absolute

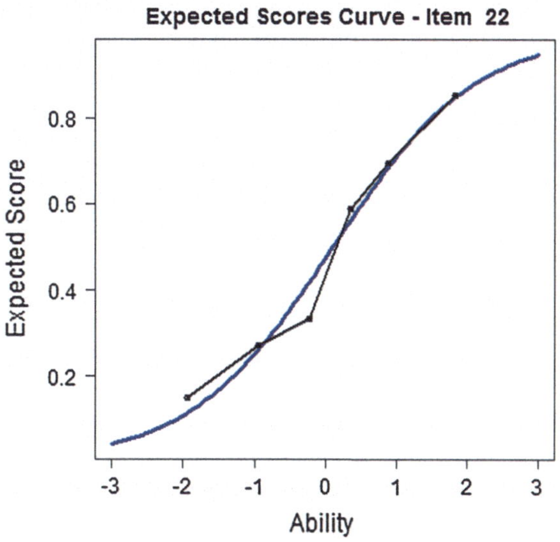

Fig. 8.4 Points of observed ICC are close to the theoretical ICC (infit MNSQ = 1.01)

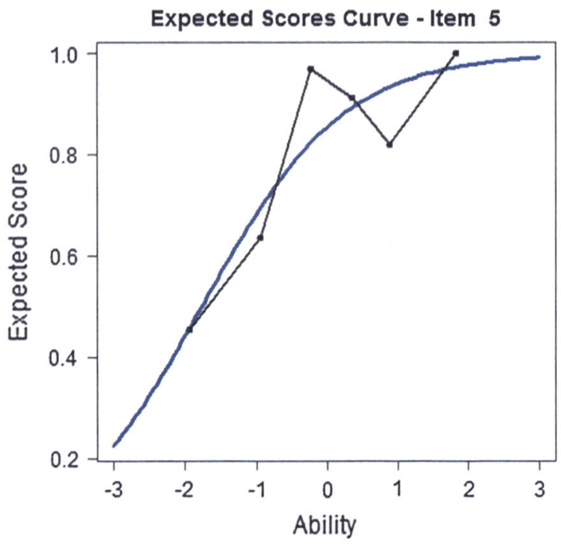

Fig. 8.5 Points of observed ICC "far away" from the theoretical ICC (infit MNSQ = 1.00)

fit". An item showing misfit in one test may very well fit with items in another test. We expand this point in the latter part of this chapter.

Discrete Observations and Fit

Figures 8.4 and 8.5 demonstrate that the visual impression of the closeness between the observed and expected ICC does not necessarily reflect the fit of an item to the Rasch model. This, in part, is due to the discreteness of the response data, namely, 0 and 1, in the case of dichotomous data. While the expected score is a number *between* 0 and 1, the observed score is either 0 or 1. If we plot the expected scores curve together with the actual observed data, we see that the observed data is nearly always "far away" from the expected scores curve. Figure 8.6 shows such a plot.

The circles in Fig. 8.6 show the observed responses, which are either 0 or 1. There are fewer 0's than 1's at the high end of the ability scale, and there are fewer 1's than 0's at the low end of the ability scale. To plot the observed ICC, we typically group response data into ability groups. The visual appearance of the observed ICC depends very much on the number of ability groups and the student sample size. For example, Fig. 8.7 shows the ICC for an item when there are 10 ability groups (left graph) and 6 ability groups (right graph), for the same item. These two graphs look quite differently in terms of the match between the observed data with the expected values. When there are fewer groups, the number of observations in each group is larger so the curve appears smoother. For very large samples, the curve will be smoother than for smaller samples. The difference in appearance of the observed ICCs shows that the residual-based fit statistics are not

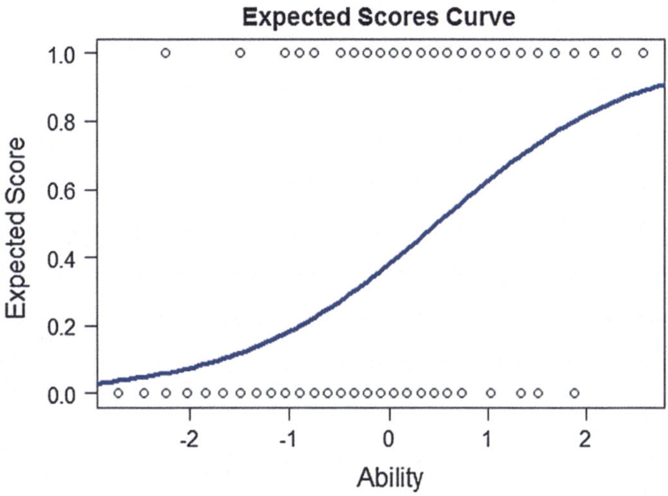

Fig. 8.6 Expected scores curve versus raw data

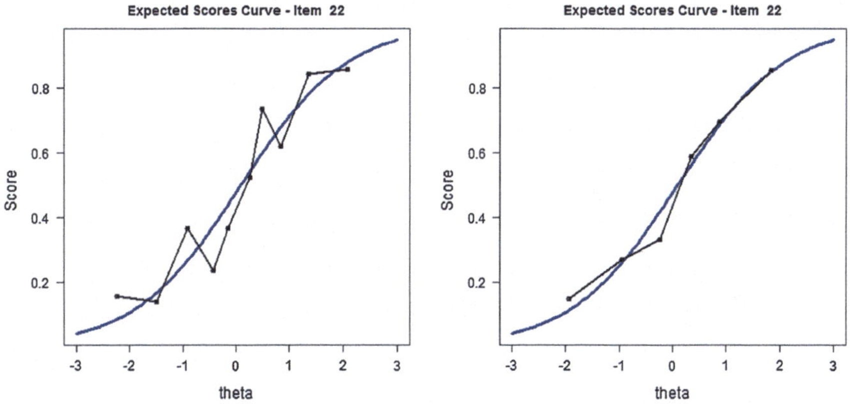

Fig. 8.7 ICCs of an item with 10 ability groups (*left*) and 6 ability groups (*right*)

the same as the concept of "goodness-of-fit" in the context of regression analysis where one checks the deviation of each observation from the expected curve.

Distributional Properties of Fit Mean-Square

In the above section about the derivation of the fit mean-square statistic (Eqs. (8.3) and (8.4)), it is stated that the expectation of these two statistics is one. That is, when the data fit the model, we expect the fit mean-square to be close to one. But "how close to one" is a judgment call. To assess "how close to one is close enough", we will need to know the amount of variation of the mean-square statistics. More formally, it can be shown that the asymptotic variance of the fit mean-square is given by $2/N$, where N is the sample size of students (see Additional Notes). This means that if a test is given to a small group of students (i.e. N is small), we would expect the fit mean-square for each item to fluctuate quite widely around one, even when the items fit the Rasch model. For example, if the sample size is 200, we would expect 95% of the mean-square values to be between 0.8 and 1.2 (standard error = $\sqrt{\frac{2}{200}} = 0.1$). In comparison, when the same test is given to a large group of students, the fit mean-square will be very close to one. For example, if the sample size is 2000, we would expect 95% of the mean-square values to be between 0.94 and 1.06 ($\sqrt{\frac{2}{2000}} = 0.03$). Since the variance of the mean-square statistic depends on the sample size, we need to be careful about applying fixed limits around one to make an assessment of the fit of an item.

Figure 8.8 shows a fit map of 20 items administered to 100 students for a simulated data set. It can be seen that the fit mean-square values are generally between 0.8 and 1.2.

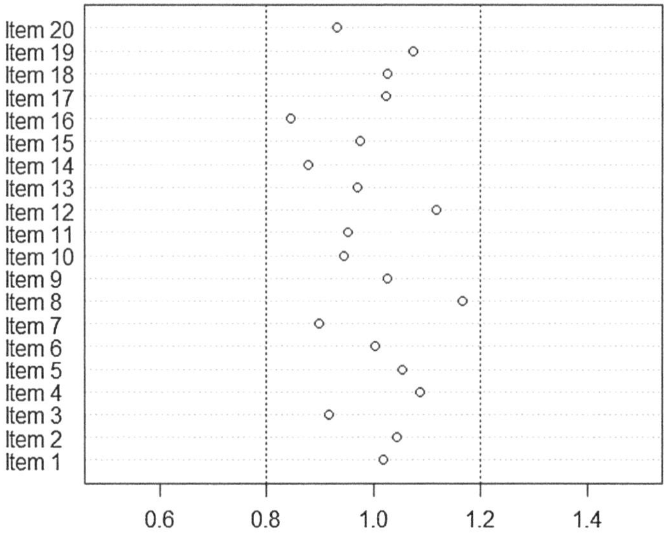

Fig. 8.8 Fit mean squares map when sample size = 100

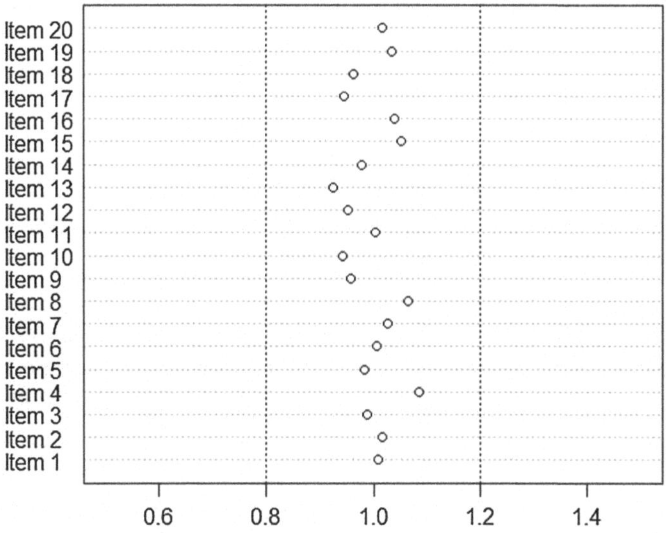

Fig. 8.9 Fit mean squares map when sample size = 500

In contrast, Fig. 8.9 shows a fit map of the same 20 items administered to 500 students. It can be seen that the fit mean-square values are generally between 0.9 and 1.10. The only difference between the two analyses is the sample size. The same items were used for both analyses. Since the data were simulated according to the Rasch model, all items were expected to fit the model. These two examples demonstrate that an assessment of the magnitude of the fit mean-square statistic should take into account of the sample size of the test administration.

Additional Notes

The numerator in the unweighted fit mean-square statistic, $\sum_n z_{ni}^2$, is an observed value of the sum of squares of random variables Z_{ni} with mean 0 and standard deviation of 1. The random variable, Z_{ni}, has a discrete distribution, as the observed response can only take values 0 and 1 (in the dichotomous case). While Z_{ni} is not a standard normal random variable, when N is large, $\sum_n Z_{ni}^2$ can be regarded as having a chi-square distribution with N degrees of freedom (note that the sum of squares of independently distributed standard normal random variables has a chi-square distribution with N degrees of freedom.) The mean of a chi-square distribution with N degrees of freedom is N, and the variance is 2N. Consequently, the asymptotic variance of the unweighted fit mean-square is $Var\left(\frac{\sum_n Z_{ni}^2}{N}\right) = \frac{1}{N^2} \times 2N = \frac{2}{N}$

The Fit *t* Statistic

The fit *t* statistic, however, does take sample size into account. Even though it is called a *t* statistic, the fit *t* statistic can be regarded as a normal deviate with a mean of zero and a standard deviation of one (i.e., a "z" score), as the sample is typically large enough to use the normal approximation. The fit *t* statistic is a transformation of the fit mean-square value, taking into account of the mean and variance of the fit mean-square statistic.

Additional Notes

To transform the fit mean-squares to a standardised normal statistic so that one can look up the level of significance easily, the Wilson-Hilferty transformation $t_{unwtt} = \left(Fit_{unwtt}^{1/3} - 1 + 2/(9N)\right)/(2/(9N))^{1/2}$ is often used, where *Fit* is the mean-square value.

An alternative transformation is given in Wright and Masters (1982) that uses a cube root transformation of the fit mean-square and its variance:

$$t_{unwtt} = \left[Fit_{unwt}^{1/3} - 1\right] \times \frac{3}{\sqrt{Var(Fit_{unwtt})}} + \frac{\sqrt{Var(Fit_{unwt})}}{3}$$

Since the fit *t* statistic can be regarded as a normal deviate, a *t* value outside the range of −2.0 to 2.0 (or −1.96 to 1.96, to be more precise) can be regarded as an indication of misfit, at the 95% confidence level.

On the surface, our problem regarding the lack of a stable frame of reference for the fit mean-square values seems to have been solved. Unfortunately, this is not the case.

The problem is, in real-life, no item fits the Rasch model perfectly. When items do not fit the Rasch model, any misfit, however small, can be detected when the sample size is large enough. This means that the fit *t* values will invariably show significance when the sample size is very large. In some sense, the *t* values are telling the "truth", that there are indeed differences between items, and the items do not tap into the same construct. However, some of these differences between items may be minute from a practical point of view.

The following shows an example of how sample size affects the fit *t* values.

Item response data from the First International Mathematics Study (FIMS) (IEA study conducted in 1964) for Australia and Japan were scaled using the Rasch model, first selecting just 500 students at random, and then selecting 2000 students at random. Finally the full data set with 6371 students was analysed. That is, the items scaled in all three samples were exactly the same, but the sample analysed increased in size. Figures 8.10, 8.11, 8.12 show the fit *t* values for these three samples.

	outfitItem	outfitItem_t	infitItem	infitItem_t
M1PTI1	1.0176412	0.1681252	0.9604863	-0.6563752
M1PTI2	0.8600060	-0.8820403	0.9132830	-1.4011322
M1PTI3	0.8747212	-0.5301774	0.9267337	-0.8653759
M1PTI6	0.8348506	-1.6447400	0.9073273	-2.1574356
M1PTI7	0.5103849	-2.8934127	0.7318945	-3.4068671
M1PTI11	0.7181897	-1.5328399	0.8410477	-2.1687957
M1PTI12	1.2043202	2.0302249	1.1212745	2.5738491
M1PTI14	1.4884776	4.9042209	1.2516368	5.6743577
M1PTI17	1.1407154	1.0991002	0.9672130	-0.5461018
M1PTI18	0.9293927	-0.6623501	0.9005372	-2.3313171
M1PTI19	0.6514076	-2.6648368	0.7768166	-3.6387173
M1PTI21	1.8352155	4.5175900	1.3942578	5.2328157
M1PTI22	1.2055754	1.1699414	1.0063787	0.1092650
M1PTI23	0.9120535	-0.6515327	0.9760430	-0.4367405

Fig. 8.10 Fit *t* values for a sample of 500 students

	outfitItem	outfitItem_t	infitItem	infitItem_t
M1PTI1	1.0617523	0.7379272	1.0214576	0.6765263
M1PTI2	0.7068081	-3.8128545	0.8672650	-4.3360205
M1PTI3	0.8524912	-1.3102948	0.9423666	-1.3888542
M1PTI6	0.8184879	-4.0480032	0.8627173	-6.9696935
M1PTI7	0.7469162	-2.9795967	0.8325752	-4.2399000
M1PTI11	0.7920345	-2.4043510	0.9034710	-2.9044195
M1PTI12	1.4255288	8.0500594	1.2021082	8.1901920
M1PTI14	1.3671890	7.4344970	1.2199722	9.3759386
M1PTI17	1.1176319	1.7561847	0.9639025	-1.1589724
M1PTI18	0.9377125	-1.2471293	0.9689911	-1.4644269
M1PTI19	0.5785473	-6.2854536	0.7602613	-7.1088021
M1PTI21	1.7399073	8.8961978	1.3602378	9.9630739
M1PTI22	0.9774774	-0.2471895	0.9211786	-2.1140544
M1PTI23	0.7957022	-3.3884918	0.9086099	-3.7092765

Fig. 8.11 Fit *t* values for a sample of 2000 students

From Figs. 8.10, 8.11, 8.12, it can be seen that as sample size increases, the fit *t* values became progressively far away from zero so that many items showed statistically significant misfit.

Item Fit Is Relative, Not Absolute

As we have mentioned, the fit mean-square statistic defined in Eq. (8.3) tests whether the item has the same "slope" as the other items in the test, since the Rasch model makes the assumption that all items have the same discrimination. The Rasch model does not, however, stipulate what the discrimination should be.

	outfitItem	outfitItem_t	infitItem	infitItem_t
M1PTI1	1.0772154	1.59714318	1.0182028	1.012313
M1PTI2	0.7127304	-7.01014880	0.8631072	-8.290753
M1PTI3	0.8548437	-2.31343680	0.9401009	-2.554359
M1PTI6	0.8511479	-5.89885727	0.8890838	-9.915885
M1PTI7	0.7157490	-5.61754107	0.8225674	-7.870286
M1PTI11	0.7980984	-4.14518615	0.8942556	-5.626692
M1PTI12	1.3649289	11.26997571	1.1814339	12.419531
M1PTI14	1.3721131	12.92338691	1.2133264	16.153633
M1PTI17	1.0984786	2.55868638	0.9547443	-2.676895
M1PTI18	1.0018899	0.07510419	0.9785618	-1.767476
M1PTI19	0.6063701	-9.90033565	0.7778577	-11.743530
M1PTI21	1.7837242	15.58244351	1.3510149	17.137759
M1PTI22	0.9783801	-0.41760535	0.9138654	-4.053861
M1PTI23	0.8246742	-5.15048395	0.9192772	-5.788562

Fig. 8.12 Fit t values for a sample of 6371 students

Consequently, items in a test will show good fit (i.e., fit mean-squares around 1) if the items have similar discrimination, even if the discrimination power is poor. That is, if all items are *equally* "bad" (here we use the term "bad" to indicate low discrimination power), the items will still show good fit, because they have equal discrimination. Consequently, when there is no mis-fitting item, we might conclude that the response data fit the Rasch model, we cannot conclude that we have the *best* test. The test reliability may still be low. Figure 8.13 shows a comparison of weighted fit mean-squares and test reliability between two 20-item tests.

The fit mean-squares of both tests show that there is no mis-fitting item, but the two tests have quite different test reliability.

In the extreme case, if every student randomly guesses answers to all questions, the items will still fit the Rasch model (items are equally (non-)discriminating). But the test reliability will be close to zero. Consequently, to check whether the test instrument as a whole has the capacity to separate students in terms of their abilities, the reliability index is still a better measure, not the fit statistics. The classical test theory discrimination index is also a good indicator of item discrimination.

Since the fit statistics test whether the items have equal discrimination, an item showing mis-fit in a test may very well show good fit in another test. For example, an "over-fit" item (fit mean-square less than 1) shows that the item is more discriminating than most other items in the test. If we take all highly discriminating items (fit mean-squares less than 1) in a test and re-run the item analysis, we will find that some of the items will now show "under-fit" and others will show "over-fit", since the fit statistics show the relative discrimination powers of items within an item set.

In practice, to select items from a trial analysis, it is best *not* to choose the set of items with fit statistics around 1 (the best fitting items), since these "good-fit"' items are items with "mediocre" discrimination. We recommend selecting the "over-fit" items (with fit mean-squares less than 1). These items are highly discriminating items.

Fig. 8.13 Comparison of fit MNSQ of two 20-item tests

	Test 1	Test 2
Item	Fit MNSQ	Fit MNSQ
I1	1.00	1.05
I2	1.03	1.02
I3	1.01	1.02
I4	0.99	0.97
I5	1.01	1.01
I6	0.99	0.97
I7	0.95	0.93
I8	1.00	0.99
I9	0.98	0.97
I10	1.00	1.01
I11	1.02	1.01
I12	1.03	1.05
I13	0.98	0.98
I14	1.05	1.10
I15	1.00	0.94
I16	1.02	1.00
I17	0.96	0.97
I18	0.99	1.03
I19	1.00	0.94
I20	1.00	1.03
Reliability	0.49	0.72

To be more explicit, suppose a set of 40 items have fit mean-squares ranging between 0.7 and 1.3. Twenty items are selected. Set A consists of items with fit mean-squares closest to 1 (say, between 0.9 and 1.1). Set B consists of items with fit mean-squares less than 1 (say, between 0.7 and 1). Both sets of items, when the item analysis is re-run, will show item fit statistics around 1 (since it's all relative to the items in the set). But Set B will have a higher test reliability than Set A.

Summary

To use fit statistics in item analysis, we need to understand the properties of these statistics. In particular, the impact of the sample size on the fit statistics needs to be taken into account. If we use fit mean-square values to set criteria for accepting or rejecting items on the basis of fit, we are likely to declare that all items fit well when the sample size is large enough. On the other hand, if we set limits to fit t values as a criterion for detecting misfit, we are likely to reject most items when the sample size is large enough.

Fig. 8.14 Expected ICC and observed ICC points

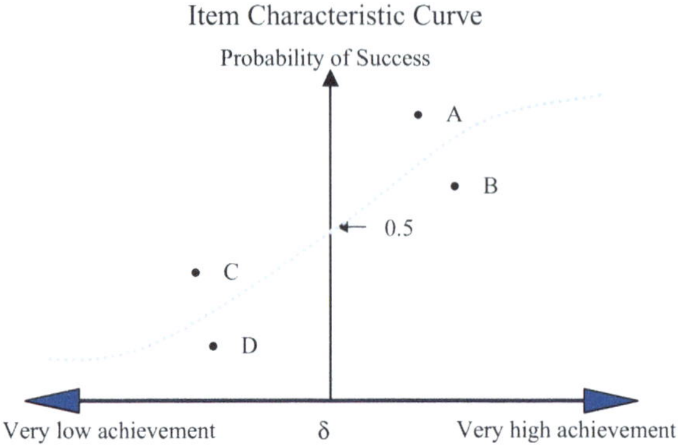

Some textbooks or other resources make recommendations on the range of acceptable mean-square values or *t* values for residual-based fit statistics. There are probably no right or wrong answers. You will need to understand the issues with these fit statistics when you apply rules of thumb.

More importantly, fit statistics should serve as an indication for detecting problematic items rather than for setting concrete rules for accepting or rejecting items. Based on the fit statistics, one should examine the items and look for sources of misfit. Improve or reject items if sources of misfit can be identified. The fit statistics should not be used blindly to reject items, particularly those that "over-fit", as you may remove the best items in your test because the rest of the items are not as "good" as these items.

Furthermore, when residual-based fit statistics show that items fit the Rasch model, this is not sufficient to conclude that you have the best test. The reliability of the test and item discrimination indices should also be considered in making an overall assessment.

Additional Notes

Figure 8.14 shows the theoretical, or expected, item characteristic curve for an item, with four points, A, B, C, and D denoting four regions where the observed ICC may fall. Point A denotes the region above the theoretical ICC, and to the right of the vertical line where $\theta = \delta$, the ability at which there is a 50% chance of obtaining the correct answer. Point B denotes the region below the theoretical ICC and to the right of the vertical line $\theta = \delta$. Point C denotes the region above the theoretical ICC but to the left of the $\theta = \delta$ line. Point D denotes the region below the theoretical ICC and to the left of the $\theta = \delta$ line. It can be shown mathematically that the contribution of observed points in the A and D region to the outfit mean-square, $z_{ni}^2 = \frac{(x_{ni} - E(X_{ni}))^2}{(Var(X_{ni}))}$, has an expectation less than one, while the expectation of z_{ni}^2 for points in the C and B regions is greater than one. It is clear then the fit mean-square value provides a test of whether the "slope" of the observed ICC is the same as the theoretical one. Given that the theoretical one can be regarded as an "average" of all items, the fit mean-square value tests whether the observed ICC for this item is the same as the slopes of the other items.

When residual-based fit statistics show that items fit the Rasch model, this is not sufficient to conclude that you have the best test instrument.

Discussion Points

(1) Explain why we say that when the item difficulty of an item is close to the ability of a person, the corresponding $Var(X_{ni})$ is relatively large?

(2) Consider Fig. 8.6 that shows the expected scores curve versus the observed responses, which are either 0 or 1, of a typical item. Explain why there are fewer 0's than 1's at the high end of the ability scale, and fewer 1's than 0's at the low end of the ability scale.

(3) Discuss what you would do after you have detected some items with fit mean squares statistics quite different from the value of 1? Should you simply label them as bad items? Are there other considerations?

(4) We explain in this chapter that with a larger sample size, the mean-square statistics of the items will be closer to 1. In a sense, the items could be considered to "fit" better under a larger sample size. On the other hand, we also point out that with a larger sample size, more items will be identified to deviate from 0 in terms of the fit t statistics. In a sense, more items are considered to "fit" worse under a larger sample size. Explain this apparent dilemma and why a larger sample size seems to have different effect in terms of different statistics. Understand how to navigate between these two types of statistics in practical works in assessment. It is important to remember that the fit statistics should not be used blindly to reject items.

Exercises

Q1. In TIMSS 2011 student questionnaire for New Zealand Year 9 students, there is a question about home possession, as show in Fig. 8.15 (TIMSS 2010). These questions could be measuring the "family wealth" construct. The data set was downloaded from the TIMSS and PIRLS website.

(*Source* TIMSS 2011 Assessment. Copyright © 2013 International Association for the Evaluation of Educational Achievement (IEA). Publisher: TIMSS & PIRLS International Study Center, Lynch School of Education, Boston College, Chestnut Hill, MA and International Association for the Evaluation of Educational Achievement (IEA), IEA Secretariat, Amsterdam, the Netherlands).

5

Do you have any of these things at your home?

Tick one circle for each line.

	Yes	No
a) Computer	○	○
b) Study desk/table for your use	○	○
c) Your own books (do not count your school books)	○	○
d) Your own room	○	○
e) Internet connection	○	○
f) Musical instruments (e.g.. piano. violin. guitar)	○	○
g) Clothes dryer	○	○
h) Dishwasher	○	○
i) Two or more bathrooms	○	○
j) Your own computer or laptop	○	○
k) Swimming pool or spa pool (do not include paddling pools)	○	○

Fig. 8.15 TIMSS student questionnaire on home possession (TIMSS 2010)

A Rasch analysis was run. Table 8.1 shows summary results of the item analysis. Based on the item analysis, discuss how well the items measure the construct of wealth.

Table 8.1 Summary of item analysis for home possession items

	Percent answering "yes"	Item difficulty δ	CTT discrimination	Infit mean squares	Infit t
Q5a	0.96	−3.74	0.30	0.93	−1.14
Q5b	0.86	−2.15	0.26	1.01	0.45
Q5c	0.88	−2.34	0.20	1.05	1.80
Q5d	0.90	−2.57	0.22	1.02	0.74
Q5e	0.91	−2.70	0.38	0.91	−2.39
Q5f	0.68	−0.93	0.20	1.10	6.54
Q5g	0.82	−1.83	0.27	1.03	1.15
Q5h	0.72	−1.12	0.45	0.90	−6.32
Q5i	0.56	−0.27	0.40	0.96	−3.51
Q5j	0.41	0.43	0.19	1.09	7.40
Q5k	0.23	1.48	0.24	1.00	−0.02

Person separation reliability: 0.47
Sample size of students: 5233

References

Andersen EB (1973) A goodness of fit test for the Rasch model. Psychometrika 38:123–140

Chen W-H, Thissen D (1997) Local dependence indexes for item pairs using item response theory. J Educ Behav Stat 22:265–289

Douglas G (1982) Issues in the fit of data to psychometric models. Educ Res Perspect 9:32–43

Glas CAW, Verhelst ND (1995) Testing the Rasch Model. In: Fischer G, Molenaar I (eds) Rasch models. Springer-Verlag, pp 69–96

Hattie J (1985) Methodology review: assessing unidimensionality of tests and items. Appl Psychol Meas 9:139–164

Kiefer T, Robitzsch A, Wu M (2013) TAM (Test Analysis Modules)—an R package [computer software]

Liu Y, Maydeu-Olivares A (2013) Local dependence diagnostics in IRT modeling of binary data. Educ Psychol Measur 73:254–274. doi:10.1177/0013164412453841

Linacre JM, Wright BD (2000) WINSTEPS: a Rasch computer program. MESA Press, Chicago

Maydeu-Olivares A (2013) Goodness-of-fit assessment of item response theory models. Measurement 11:71–101

McDonald R, Mok MM-C (1995) Goodness of fit in item response models. Multivar Behav Res 30 (1):23–40

Meijer RR, Sijtsma K (2001) Methodology review: evaluating person fit. Appl Psychol Measur 25:107–135

Molenaar IW (1983) Some improved diagnostics for failure of the Rasch model. Psychometrika 48:49–72

Laboratory RUMM (2001) Rasch unified measurement models. Author, Perth

Smith RM (1988) The distributional properties of Rasch standardized residuals. Educ Psychol Measur 48:657–667

Stout W, Habing B, Douglas J, Kim HR, Roussos L, Zhang J (1996) Conditional covariance-based nonparametric multidimensionality assessment. Appl Psychol Measur 20(4):331–354

TIMSS (2010) TIMSS 2010/2011 student questionnaire. TIMSS & PIRLS International Study Center, Lynch School of Education, Boston College, in conjunction with New Zealand TIMSS

National Research Centre, Wellington, New Zealand. Retrieved 29 July 2014 from http://www.educationcounts.govt.nz/topics/research/timss/timss_1011

van den Wollenberg AL (1982) Two new test statistics for the Rasch model. Psychometrika 47:123–139

Wright BD (1977) Solving measurement problems with the Rasch Model. J Educ Measur 14:97–116

Wright BD, Masters GN (1982) Rating scale analysis. MESA Press, Chicago

Wright BD, Panchapakesan N (1969) A procedure for sample-free item analysis. Educ Psychol Measur 29:23–48

Wu ML, Adams RJ (2013) Properties of Rasch residual fit statistics. J Appl Measur 14(4):339–355

Wu ML, Adams RJ, Wilson MR (1998) ConQuest: generalised item response modelling software. Australian Council for Educational Research, Camberwell

Chapter 9
Partial Credit Model

Introduction

For some measuring instruments, item responses may reflect a degree of correctness (or a degree of appropriateness in the case of survey questionnaires) in the answer to a question, rather than being simply classified as correct/incorrect. To model these item responses, the Partial Credit Model (PCM) (Masters 1982) can be applied where item scores have more than two ordered categories (also known as polytomous items).

The partial credit model has been applied to a wide range of item types. Some examples include the following:

1. Likert type questionnaire items that allow options such as strongly agree, agree, disagree, strongly disagree.
2. Essay ratings, for example, on a scale from 0 to 5.
3. Items requiring composite processes, such as a problem-solving item requiring students to perform multiple-step procedures including formulating the problem and carrying out computation.
4. Items where some answers are more correct than others. For example, if one is asked to name the capital city of Australia, then "Sydney" is a better answer than "Auckland", even though both are incorrect.
5. A "testlet" or "item bundle" consisting of a number of questions relating to one stimulus. The total number correct for the testlet is sometimes modelled with the PCM.

Are all of the above item types appropriate for applying the PCM? How does one interpret the PCM item parameters in relation to the different item types?

Further, there are a number of different ways for the parameterisation of PCM, and for constructing measures of "difficulty" in relation to a partial credit item. A clear understanding of the "item difficulty" parameters in PCM is important when described proficiency scales (or learning progressions) are constructed. The skills

M. Wu et al., *Educational Measurement for Applied Researchers*,
DOI 10.1007/978-981-10-3302-5_9

descriptions along the ability scale are associated with the "item locations" on the scale.

The Derivation of the Partial Credit Model

It will be helpful to first describe the derivation of the PCM to clarify the underlying assumptions of a PCM. Masters (1982) derived the PCM by applying the dichotomous Rasch model to adjacent pairs of score categories. That is, given that a student's score is k-1 or k, the probability of being in score category k rather than in category k-1 has the form of the simple Rasch model.

Consider a 3-category partial credit item, with 0, 1 and 2 as possible scores for the item.

The PCM specifies that, while conditioning on scoring a 0 or 1 (i.e., we know the score is either 0 or 1), the probability of a score of zero ($X = 0$) and the probability of a score of 1 ($X = 1$) are given by

$$\begin{aligned} p_{0/0,1} &= \Pr(X = 0/X = 0 \quad \text{or} \quad X = 1) = \frac{\Pr(X = 0)}{\Pr(X = 0) + \Pr(X = 1)} \\ &= \frac{1}{1 + \exp(\theta - \delta_1)} \end{aligned} \tag{9.1}$$

$$\begin{aligned} p_{1/0,1} &= \Pr(X = 1/X = 0 \quad \text{or} \quad X = 1) = \frac{\Pr(X = 1)}{\Pr(X = 0) + \Pr(X = 1)} \\ &= \frac{\exp(\theta - \delta_1)}{1 + \exp(\theta - \delta_1)} \end{aligned} \tag{9.2}$$

Equations (9.1) and (9.2) are in the form of the dichotomous Rasch probabilities.

Similarly, conditional on scoring a 1 or 2, the probability of X = 1 and the probability of X = 2 are given by

$$\begin{aligned} p_{1/1,2} &= \Pr(X = 1/X = 1 \quad \text{or} \quad X = 2) = \frac{\Pr(X = 1)}{\Pr(X = 1) + \Pr(X = 2)} \\ &= \frac{1}{1 + \exp(\theta - \delta_2)} \end{aligned} \tag{9.3}$$

$$\begin{aligned} p_{2/1,2} &= \Pr(X = 2/X = 1 \quad \text{or} \quad X = 2) = \frac{\Pr(X = 2)}{\Pr(X = 1) + \Pr(X = 2)} \\ &= \frac{\exp(\theta - \delta_2)}{1 + \exp(\theta - \delta_2)} \end{aligned} \tag{9.4}$$

Equations (9.3) and (9.4) are also in the form of the dichotomous Rasch probabilities.

PCM Probabilities for All Response Categories

While the derivation of the PCM is based on specifying probabilities for adjacent score categories, the probability for each score, when all score categories are considered collectively, can be derived. The following gives the probability of each score category for a 3-category (0, 1, 2) PCM.

$$p_0 = \Pr(X = 0) = \frac{1}{1 + \exp(\theta - \delta_1) + \exp(2\theta - (\delta_1 + \delta_2))} \quad (9.5)$$

$$p_1 = \Pr(X = 1) = \frac{\exp(\theta - \delta_1)}{1 + \exp(\theta - \delta_1) + \exp(2\theta - (\delta_1 + \delta_2))} \quad (9.6)$$

$$p_2 = \Pr(X = 2) = \frac{\exp(2\theta - (\delta_1 + \delta_2))}{1 + \exp(\theta - \delta_1) + \exp(2\theta - (\delta_1 + \delta_2))} \quad (9.7)$$

More generally, if item i is a polytomous item with score categories 0, 1, 2, …, m_i, the probability of person n scoring x on item i is given by

$$\Pr(X_{ni} = x) = \frac{\exp \sum_{k=0}^{x} (\theta_n - \delta_{ik})}{\sum_{h=0}^{m_i} \exp \sum_{k=0}^{h} (\theta_n - \delta_{ik})} \quad (9.8)$$

where we define $\exp \sum_{k=0}^{0} (\theta_n - \delta_{ik}) = 1$, and hence when the score is 0, the numerator of Eq. (9.8) is 1. The summation index k refers to score categories.

Note that the number of δ_k parameters is one less than the number of response categories. For example, if there are three response categories, 0, 1 and 2, then there are two δ parameters, δ_1 and δ_2. In the same way as for dichotomous items, when there are two response categories (e.g., correct and incorrect), there is one item difficulty parameter, δ.

Some Observations

Dichotomous Rasch Model Is a Special Case

Note that the simple dichotomous Rasch model is a special case of the PCM. That is, when an item has two response categories (dichotomous), Eq. (9.8) is the dichotomous Rasch model as shown in Chap. 7. For this reason, software programs that can fit the PCM can generally fit the dichotomous model without special instructions to distinguish between the dichotomous model and PCM. Dichotomous and partial credit items can generally be "mixed" in one analysis.

The Score Categories of PCM Are "Ordered"

The score categories 0, 1, 2, …, m, of a PCM item should be "ordered" to reflect increasing competence of some trait. Under the PCM, there is an assumption that students with higher abilities are more likely to score higher for the item.

Consider the lowest two score categories: 0 and 1. Since the simple dichotomous Rasch model applies if we consider the case where the score categories are only 0 and 1, then students with higher abilities are more likely to achieve a score of 1 than 0. By the same token, if we consider scores 1 and 2, then higher ability students are more likely to achieve a score of 2 than 1. Consequently, when we consider all score categories for a partial credit item, higher ability students are expected to score higher than lower ability students. That is, increasing scores within an item should reflect increasing difficulty of the task.

PCM Is not a Sequential Steps Model

The derivation of PCM simply specifies the "conditional probability" of two adjacent score categories. The PCM does not make any assumption that there is an underlying sequential step process to achieve a score. That is, there is no assumption that a student must be successful in *all* tasks for lower score categories to achieve success in tasks for a higher score. In fact, the Steps model (Verhelst et al. 1997) should be used for items where students cannot achieve a higher score unless tasks for lower scores are successfully completed (a sequential step process).

This observation is important for the interpretation of the item parameters, δ_k. In the above example where there are 3 score categories, the parameter, δ_2, does not reflect the item difficulty of being successful in both "steps", or for achieving a score of 2. Nor does δ_2 reflect the item difficulty for the second "step" as an *independent* item.

It is worth noting that the IRT models considered here are probabilistic and not deterministic. That is, if a student obtains a score of 2 on an item, it does *not* mean that it is with certainty that the student can performed the task demands for getting a score of 1. Under a probabilistic model, the student is *likely* to perform the task demands of a score of 1, but there is always a chance that the student cannot perform all the task demands for a score of 1.

The Interpretation of δ_k

The derivation of the PCM, based on the simple Rasch model for adjacent score categories, leads to the misconception that δ_k is the difficulty parameter for step k, should step k be administered as an independent item. The interpretation of δ_k can be clarified graphically through the item characteristic curves.

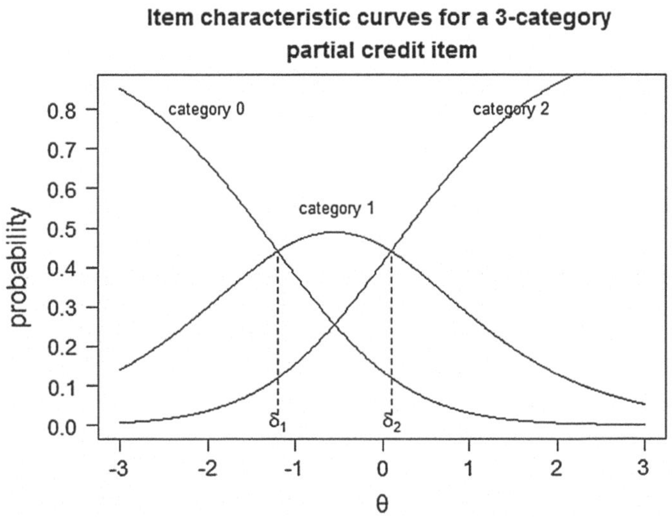

Fig. 9.1 Theoretical item characteristic curves for a 3-category partial credit Item

Item Characteristic Curves (ICC) for PCM

Item characteristic curves for a partial credit item are plots of the probabilities of being in each score category, as a function of the ability, θ. Figure 9.1 shows example item characteristic curves for a 3-category partial credit item.

From Fig. 9.1, it can be seen that as ability increases, the probability of being in a higher score category also increases.

Graphical Interpretation of the Delta (δ) Parameters

Mathematically, it can be shown that the delta (δ) parameters in Eqs. (9.1)–(9.4) are the abilities at which adjacent ICCs intersect. That is, δ_k is the point at which the probability of being in category k − 1 and category k is equal.[1] This mathematical fact provides an interpretation for the delta (δ) parameters. Figure 9.1 shows the ICCs of a 3-category partial credit item. It can be seen that the two delta parameters, δ_1 and δ_2, divide the ability continuum into three regions. From $-\infty$ to δ_1, the most likely single score category is “0”, because the Pr(X = 0) curve is higher than either of the Pr(X = 1) curve or the Pr(X = 2) curve. Between δ_1 and δ_2, the most likely single score category is “1”. When the ability of a student is above δ_2, the most likely single score category is “2”.

[1]This probability is not 0.5, but less than 0.5, because the probability of being in categories other than k − 1 and k is not zero.

The phrase "the most likely single score category" is used to stress that it is the most likely score category at an ability level when each individual score category is considered. For example, in Fig. 9.1, between δ_1 and δ_2, score 1 has a higher probability than either score 0 or score 2. However, the combined probability of scores 0 and 2 is higher than the probability of score 1. Since the probability of score 1 is less than 0.5 between δ_1 and δ_2, so the combined probability of scores 0 and 2 is more than 0.5 for this example item.

Consequently, if the delta (δ) parameters are used as indicators of "item difficulty", one might say that δ_1 is a point such that for students with abilities beyond this point, the probability of achieving a score of 1 is higher than the probability of achieving a score of 0. Similarly, when ability is higher than δ_2, the probability of achieving a score of 2 is higher than the probability of achieving a score of 1. But we stress that the probabilities at these points are not 0.5, as for the dichotomous case.

Problems with the Interpretation of the Delta (δ) Parameters

For some items, the interpretations of the delta parameters may become problematic as the delta (δ) parameters may not be ordered. Figure 9.2 shows an example.

Figure 9.2 shows that the probability curve for the middle category, score 1, is very *flat*, indicating that there are few students who are likely to score 1. One might say that score 1 is not a very "popular" category. In this case, the interpretation of the ICCs becomes more difficult, as score 1 is never the most likely single category for any ability level, and the parameters δ_1 and δ_2 are not ordered ($\delta_1 > \delta_2$). However, δ_1 is still the ability at which the probability of being in category 1 exceeds the probability of being in category 0, and δ_2 is the ability at which the

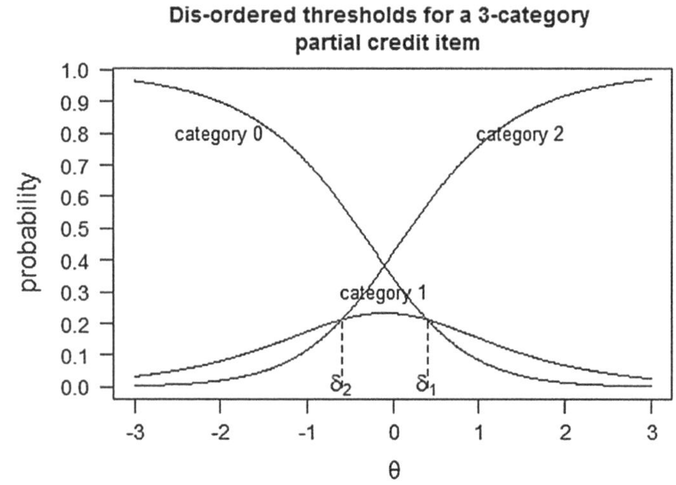

Fig. 9.2 ICC for PCM where the delta parameters are dis-ordered

probability of being in category 2 exceeds the probability of being in category 1. Since the derivation of the PCM involves specifying the probabilities for pairs of adjacent response categories, the PCM does not require the specification of the order (or prevent the dis-ordering) of the δ parameters. In other words, the dis-ordering of δ parameters does not violate any assumptions of the PCM.

Nevertheless, this phenomenon of dis-ordering of the δ_k parameters is one disadvantage of using the delta (δ) parameters to interpret item responses in relation to ability. Adams et al. (2012) provide an in-depth discussion on the issues of dis-ordered thresholds.

Linking the Graphical Interpretation of δ to the Derivation of PCM

Masters and Wright (1997) pointed out that the dis-ordering of the delta (δ) parameters was not necessarily an indication of a problematic item, since the derivation of the partial credit model did not place any restriction on the ordering of item parameters, δ. When all score categories are considered in an ICC plot such as those shown in Figs. 9.1 and 9.2, the δ parameter is the value at which adjacent score categories have equal probability. However, the probability is no longer 0.5, since there is the possibility of being in score categories other than k-1 and k. It can be seen from Figs. 9.1 and 9.2 that the point of intersection of two adjacent categories will depend on the relative chances of being in all categories. For example, in Fig. 9.2, if the probability of being in category 1 is small throughout the whole ability range (may be due to an easy step "2"), then the point of intersection (equal probability) between category 0 and 1 is likely to be at a high ability value, and the intersection point between category 1 and 2 is likely to be at a low ability value. It is clear then that the delta (δ) parameters are dependent on the number of students in each score category, so δ cannot reflect "independent" step difficulty. Rather, the values of δ will depend on the difficulties of all "steps". See Verhelst and Verstralen (1997) for an example about the dependence between the delta (δ) parameters.

Examples of Delta (δ) Parameters and Item Response Categories

When the PCM is applied to items where score categories correspond to sequential "steps" to solve a problem, the problem of dis-ordering of δ is likely to occur. This is because that, very often, latter steps are easy steps as compared to earlier steps. For example, a mathematics item involving a first step of conceptualising the method (score category 1) and a second step of carrying out computation (score category 2) will often result in most students being in either the 0 category or the 2 category. That is, few students who successfully conceptualised the method will make a computational mistake. As an example, Fig. 9.3 shows a mathematics word problem that requires formulation of an equation and then carrying out computation to obtain the result. The item statistics in Fig. 9.3 show that only 4% of students who used the correct method but made a computational error (score of 1). Figure 9.4 shows the corresponding item characteristic curves. Category 1 curve is very flat as very few students are in this response category, so that the probability of being in this response category is very low. As category 1 is never the most likely response category across the ability range, dis-ordered δ occurs ($\delta_1 > \delta_2$). In this example, δ_1 is 1.85 and δ_2 is -2.69.

Notice that dis-ordering of the thresholds indicates that a middle response category has few respondents. This in itself is not an indication that the response categories ought to be combined. In later sections of this chapter, the issue of

Item 5 - pharm

In the Pharmochem company, there are 57 employees. Each employee speaks either German or English, or both. 25 employees can speak German and 48 employees can speak English. How many employees can speak both German and English? Show how you found your answer.

Item analysis (Item 5 – pharm)

Response	Score	Count	% of tot	Pt Bis
16*	2	293	61.68	0.43
comp err	1	18	3.79	0.01
Other	0	117	24.63	-0.36

Scoring guide:
Fully correct answer was given a score of 2. For responses with correct method but incorrect computation, a score of 1 was awarded.

*Correct answer

Fig. 9.3 Item statistics for a partial credit scoring mathematics item

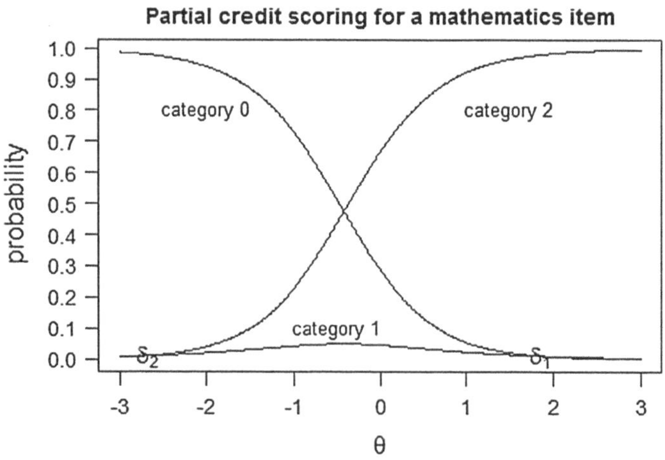

Fig. 9.4 ICC for a partial credit mathematics item with dis-ordered thresholds

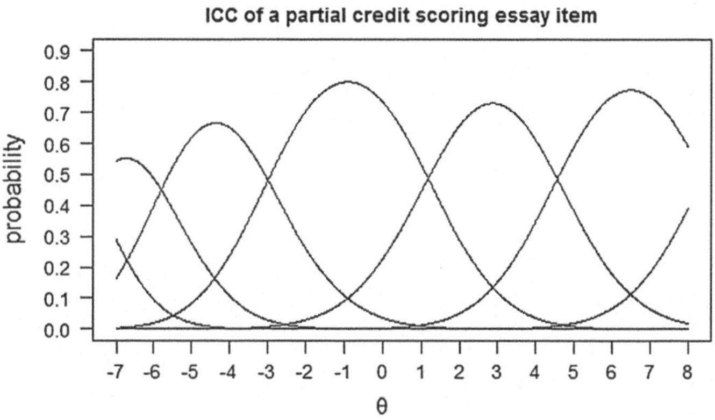

Fig. 9.5 ICC for an essay marking criterion, "Cohesion", using PCM on a 7-point scale

collapsing categories is discussed. Further, we cannot stress strongly enough that the dis-ordering of the thresholds does not mean that the scoring needs to be reversed. That is, if $\delta_1 > \delta_2$, it does *not* mean that score 1 should be labelled 2, and score 2 should be labelled 1.

In contrast to the above example, when the PCM is applied to holistic scoring rubrics such as those used for essay marking, the problem of dis-ordering of δ is less likely to occur. Figure 9.5 shows the ICC of a partial credit scoring essay item.

Tau's and Delta Dot

A variation of the parameterisation of the PCM is the use of τ's (tau's) and $\delta_\bullet$ (delta dot). Mathematically, the delta (δ_{ik}) parameters in Eq. (9.8) can be re-written in the following way:

Using the notations as in Eq. (9.8) but dropping the index i for simplicity, let

$$\delta_\bullet = \sum_{k=1}^{m} \delta_k / m \tag{9.9}$$

where m is the maximum score. That is, the total number of response categories of an item is $m + 1$.

Equation (9.9) shows that $\delta_\bullet$ is the average of the delta (δ_k) parameters for one item.

Next, let us define τ_k as the difference between $\delta_\bullet$ and δ_k. That is,

$$\tau_k = \delta_\bullet - \delta_k \tag{9.10}$$

Graphically, the relationships among τ_k, $\delta_\bullet$ and δ_k are illustrated in Fig. 9.6.

The parameterisation of the PCM using $\delta_\bullet$ and τ_k is mathematically equivalent to the parameterisation using δ_k. Using Eqs. (9.9) and (9.10), one can compute $\delta_\bullet$ and τ_k from δ_k. Conversely, given τ_k, and $\delta_\bullet$, one can compute δ_k as

$$\delta_k = \delta_\bullet - \tau_k \tag{9.11}$$

Interpretation of $\delta_\bullet$ and τ_k

The parameter $\delta_\bullet$ may be thought of as a kind of "average" item difficulty for a partial credit item. This may be useful, if one wishes to have *one* indicative difficulty parameter for a partial credit item as a whole. Otherwise, to describe the difficulty of a partial credit item, one needs to describe the difficulties of individual score categories within the item, such as using the Thurstonian thresholds described in the next section.

The τ_k parameters are more difficult to interpret as stand-alone values. These need to be interpreted in conjunction with $\delta_\bullet$. That is, τ_k, considered as a "step parameter", shows the distance of a partial credit score category from the "average" item difficulty. The τ_k parameters suffer from the same problem as δ_k's, in that the τ_k's can be dis-ordered.

Note that as $\delta_\bullet$ is the average of δ_k's, if we sum up both sides of Eq. (9.11) across the categories, k, we obtain

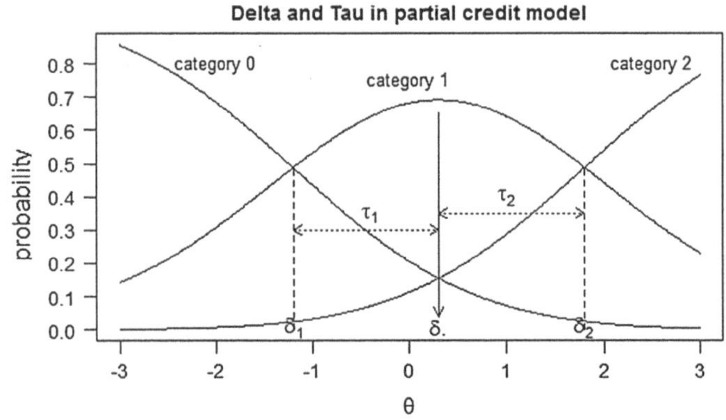

Fig. 9.6 Item characteristic curves for a three-category item with tau's and deltas

$$\begin{aligned}\sum_{k=1}^{m}\delta_k &= \sum_{k=1}^{m}(\delta_\bullet - \tau_k)\\ \sum_{k=1}^{m}\delta_k &= \sum_{k=1}^{m}\delta_\bullet - \sum_{k=1}^{m}\tau_k\\ \sum_{k=1}^{m}\delta_k &= \sum_{k=1}^{m}\frac{\sum_{k=1}^{m}\delta_k}{m} - \sum_{k=1}^{m}\tau_k\\ \sum_{k=1}^{m}\delta_k &= \sum_{k=1}^{m}\delta_k - \sum_{k=1}^{m}\tau_k\\ \sum_{k=1}^{m}\tau_k &= 0\end{aligned} \tag{9.12}$$

In the case of three response categories (0, 1, 2), we have $\tau_1 + \tau_2 = 0$ from Eq. (9.12), so $\tau_1 = -\tau_2$. In general, the sum of the τ_k parameters is zero, so there is a constraint on the τ_k parameters. In some software programs, only the first $m - 1$ τ_k parameters are estimated, and the last τ_m is set to the negative sum of the other τ_k parameters. In general, if the total number of response categories of a PCM item is K ($=m+1$), then there are K − 2 τ parameters estimated. As an example, if a PCM item has 4 categories, 0, 1, 2, 3, then, using the delta δ parameterisation, three δ parameters are estimated. Using the $\delta_\bullet$ and τ parameterisation, one $\delta_\bullet$ is estimated, and two τ parameters are estimated. In all, there are still three parameters estimated as in the case for the delta δ parameterisation.

Additional Notes

Mathematically, $\delta_\bullet$ is the intersection point of the probability curves for the first and last score categories of a partial credit item. For example, if there are 3 score categories as shown in Fig. 9.6, $\delta_\bullet$ is the intersection point of the curves for category 0 and category 2. In the case of a 3-category partial credit item, category 0 curve and category 2 curve are symmetrical about $\delta_\bullet$. That is, category 0 curve is a reflection of category 2 curve about the line $\theta = \delta_\bullet$, and category 1 curve is symmetrical about the line $\theta = \delta_\bullet$ (see Fig. 9.6). Interested readers can prove this property mathematically. However, this is not usually the case when the number of score categories is more than 3, as given by the following example for a 5-category partial credit item.

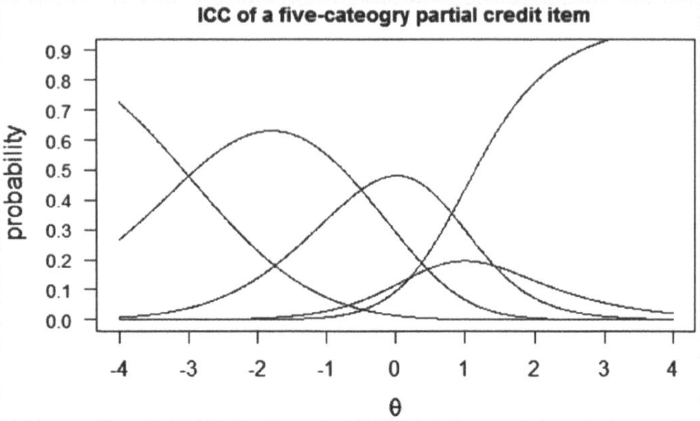

Thurstonian Thresholds, or Gammas (γ)

As was discussed in previous sections, the delta (δ) parameters do not necessarily reflect the difficulty of achieving a score point in a partial credit item. For partial credit items, to achieve a score of 2, students would generally need to accomplish more tasks than for achieving a score of 1. To reflect this "cumulative achievement", the Thurstonian thresholds are sometimes used as indicators of "score difficulties".

The Thurstonian threshold for a score category is defined as the ability at which the probability of achieving *that score or higher* reaches 0.50. Graphically, the Thurstonian thresholds are shown in Fig. 9.7.

Figure 9.7 shows the cumulative probability curves for a 5-category partial credit item. The curves show the probability of achieving a score of *1 or more*, *2 or more*, and so on. Note that the cumulative probability curve for a score of *0 or more* is just the horizontal line at the probability value of 1. Since "*0 or more*" means "any response category". The probability of this event is 1. In the example shown in Fig. 9.7, Pr(>=4) is the same as Pr(=4), since there is no response category higher than 4.

Interpretation of the Thurstonian Thresholds

Consider Fig. 9.7. Moving along the horizontal ability scale from $-\infty$ to γ_1, the probability of achieving a score of 1 or more is less than 0.5 (because the Pr(>=1)

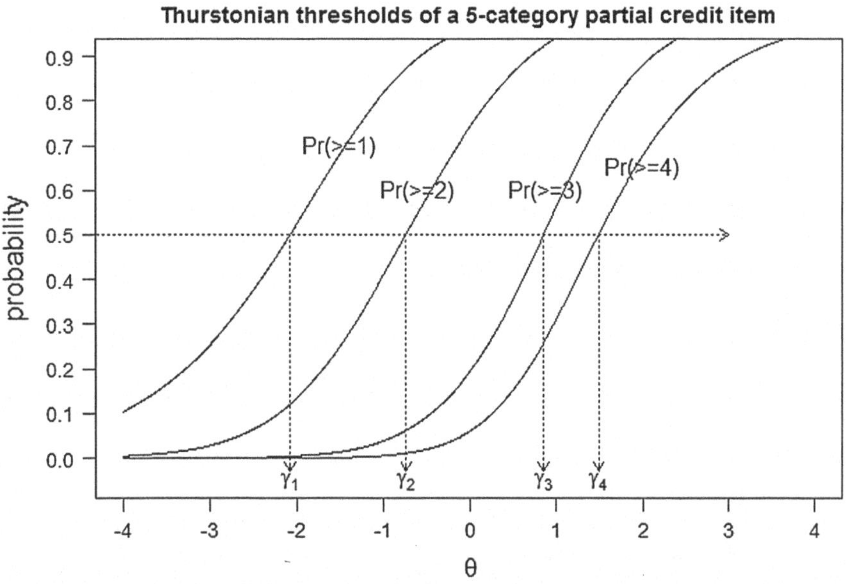

Fig. 9.7 Cumulative probability curves to show Thurstonian thresholds

curve is less than 0.5 in this range). The probability of achieving a score of 0 is more than 0.5. Therefore one might label the region from $-\infty$ to γ_1 as the "score 0" region. As the ability increases from γ_1 to γ_2, the probability of achieving a score of 1 or more is more than 0.5 (the Pr(>=1) curve), but the probability of achieving 2 or more is less than 0.5 (the Pr(>=2) curve). So one might label the region from γ_1 to γ_2 as "score 1" region. In the same manner, we can label the "score 2", "score 3" and "score 4" regions.

From this point of view, Thurstonian thresholds can be viewed as cutpoints for dividing up the ability continuum into "score regions".

So, how do Thurstonian thresholds represent item score difficulties? Is γ_1 a suitable measure for the difficulty of score 1, or is the region between γ_1 to γ_2 a better indication of score 1 "difficulty"? Should we use the mid-point between γ_1 to γ_2 as a measure of score 1 difficulty?

Comparing with the Dichotomous Case Regarding the Notion of Item Difficulty

In the dichotomous case, item difficulty is defined as the ability at which the probability of success on the item is 0.5. From this point of view, item difficulty for the dichotomous case is also a threshold, and it divides the ability continuum into two regions: score 0 and score 1 regions, and the item difficulty is the point where score 1 region starts. Extending this notion to the PCM, the Thurstonian thresholds can also be regarded as "score difficulties". That is, γ_1 is a measure of score 1 difficulty, and γ_2 is a measure of score 2 difficulty, and so on. For example, if the Thurstonian thresholds (in logits) for a 3-category item are –1.2 and 2.3, this suggests that it is relatively easy to receive a score of 1, but relatively difficult to receive a score of 2. Note that since Thurstonian thresholds are based on cumulative probabilities where P(X >= k) is always larger than (P(X >= k + 1), Thurstonian thresholds are never dis-ordered, making them more suitable for the interpretation of item difficulty for partial credit items.

Compare Thurstonian Thresholds with Delta Parameters

Dichotomous Case

As the dichotomous Rasch model is a special case of the partial credit model, the notion of Thurstonian thresholds also applies. In the dichotomous case, the Thurstonian thresholds are equal to the delta (δ) parameters.

Partial Credit Case

Depending on whether δ or γ are used as estimates of item difficulty, different difficulty measures are obtained. In the case of 3-category items, it can be shown mathematically that the Thurstonian thresholds are always "wider" than the deltas, if there are no reversals of the delta values. Figure 9.8 shows an example of comparisons between δ and γ values. Readers can prove this property as an exercise.

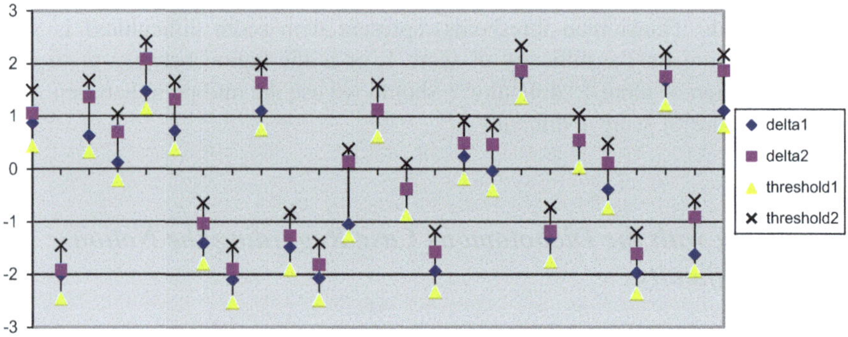

Fig. 9.8 Comparisons of threshold and delta values for 25 items

Further Note on Thurstonian Probability Curves

It should be noted that the Thurstonian probability curves such as those shown in Fig. 9.7 are not "parallel" in that the slopes of these curves are not equal. Consequently this may pose a problem when making inferences on the probabilities of the Thurstonian curves. For example, if one wants to know the ability at which there is a 75% chance of obtaining a score or higher, it is not straightforward to find the ability from a probability curve. As these curves involve cumulative probabilities, there is no analytical solution to the probability functions to solve for ability measures. Numerical methods are required to find the abilities for a given probability on the Thurstonian probability curves. In contrast, when response probability (RP) is discussed in Chapter Seven, there is a simple formula to compute the ability for any given probability when the item difficulty is known.

Using Expected Scores as Measures of Item Difficulty

Another measure of item difficulty for partial credit scoring items can be derived by computing the expected score on an item, as a function of ability. Consider an item with 3 score categories. The probabilities of scoring a 0, 1 or 2 are given by Eqs. (9.5)–(9.7). The expected score, E, on this item, as a function of the ability θ and delta parameters δ_1 and δ_2, is given by

$$E = 0 \times \Pr(X = 0) + 1 \times \Pr(X = 1) + 2 \times \Pr(X = 2) \quad (9.13)$$

Computing E as a function of θ, one can construct an Expected Scores Curve, similar to the item characteristic curve. Figure 9.9 shows an example.

Let E_1 be the ability at which the expected score on this item is 0.5. Let E_2 be the ability at which the expected score is 1.5. One might regard the region between E_1 and E_2 as the "score 1 region", and the ability continuum below E_1 as the "score 0 region", and the ability continuum above E_2 as the "score 2 region". In this way, E_1 could be regarded as an item difficulty parameter for score 1, and E_2 could be regarded as an item difficulty parameter for score 2.

The advantage of using E_1 and E_2 as indicators of difficulty is that the notion of expected scores is readily comprehensible to the layman. In the case of Thurstonian thresholds, the notion of cumulative probability is more difficulty to explain.

In addition, the expected scores curves provide a clearer comparison between the theoretical model and the observed data, in contrast to the ICCs of a partial credit model. For example, Fig. 9.10 shows the ICCs (left graph) and the expected scores curve (right graph) of an item with scores 0, 1 and 2. The expected scores curve (right graph) clearly shows that the item is not as discriminating as the model expects, but this is not so obvious from the ICCs (left graph). Consequently, to have an overview of how an item "performs" in a test, the expected scores curves are preferable.

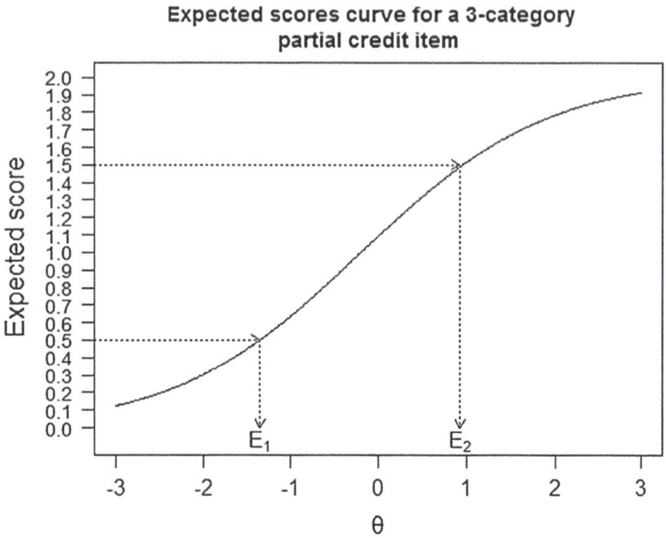

Fig. 9.9 Expected scores curve for a 3-category partial credit item

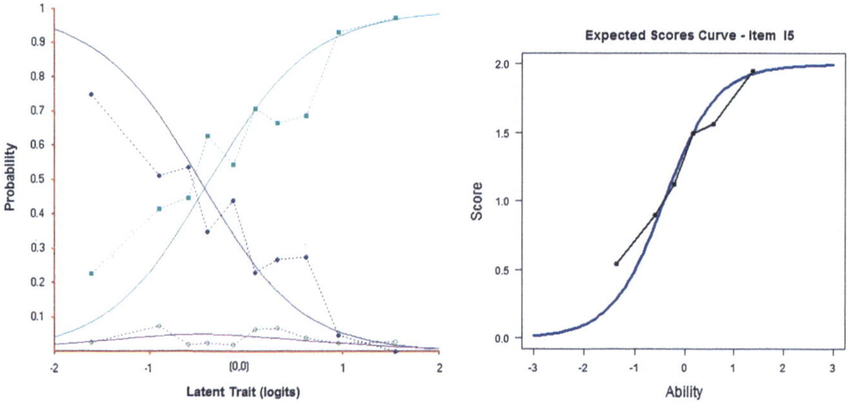

Fig. 9.10 ICCs (*left graph*) and expected scores curve (*right graph*) for a PCM item

Additional Notes

Sum of Dichotomous Items and the Partial Credit Model

Verhelst and Verstralen (1997) showed that if a set of dichotomous items fit the Rasch model, then the sum of individual item scores can be modeled using the partial credit model. However, the converse is not true. Polytomous item scores fitting the partial credit model cannot always be decomposed into individual Rasch item scores. Verhelst and Verstralen made the following statement regarding using sum scores for testlets (Note: A testlet is a set of items usually based on a common stem):

> If the main purpose of the model construction is to determine θ as accurate as possible, no information with respect to θ is lost if local independence is not violated; if it is violated, the embarrassing implications are avoided by considering sums of item scores. (p. 12)

That is, if there is a reason to think that there is dependency between a set of items, then a better way is to model the set of items as one partial credit item. The dependency will be taken into account then. However, it will not be possible to match the item parameters to individual items in the set.

Applications of the Partial Credit Model

The first part of this chapter presents the partial credit model mathematically, discusses different parameterisations and their relationships, and provides interpretations of the parameters. In the second part of this chapter, we give considerations to applications of the partial credit model, including guidelines for constructing scoring schemes and identification of mis-fitting items, illustrated with example analyses on an item.

Awarding Partial Credit Scores to Item Responses

Consider a test paper where there is a mix of multiple-choice and open-ended items. Each multiple-choice item is dichotomously scored (0 or 1), while each open-ended item has partial credit scoring (say, out of a maximum of 3 score points per item). In this case, each open-ended item is "worth" three multiple-choice items, in the sense that getting one open-ended item right is equivalent to getting three multiple-choice items right. One may query whether each open-ended item should have three times the weight of a multiple-choice item. What if the maximum score is 2 or 4? How was the maximum score of 3 decided? It is likely that it was an arbitrary decision. How does one check whether this was an appropriate decision?

First, note that if a partial credit item with a maximum score of 3 fits a PCM model, then the item with a different maximum score will not fit the PCM model. That is, we cannot arbitrarily set the maximum score of an item and expect that the item will fit the IRT model. Second, as the maximum score of an item is a weight for the item in the overall test, the magnitude of this weight should relate to how "good" this item is. A poorly functioning item in separating low and high ability students should not receive a high weight (or a high maximum score). In contrast, an item that can clearly separate students of different abilities should receive a higher weight. By this we mean that high ability students will be very likely to receive a higher score on this item, and low ability students a lower score. This is precisely the concept of item discrimination. Consequently, the weight of an item, or the maximum score of an item, should depend on the discrimination power of the item.

Very often, when a scoring guide is designed, the maximum score for an open-ended item is set based on a number of criteria which may or may not relate to the discrimination power of an item. Some examples are given below.

- A mathematics item requires 3 steps to solve it, so the maximum score is set at 3.
- The maximum score depends on the difficulty of an item. The more difficult the item, the higher the maximum score.
- An item has a number of "naturally" graded categories. For example, there could be four education levels: primary, secondary, tertiary and post-graduate. So the maximum score measuring education level is 4.

None of the above examples is the correct way of setting the maximum score of an item, since these criteria are generally unrelated to item discrimination. Take item difficulty as an example. For a multiple-choice test, we seem happy to award a score of 1 for each item. Yet the items typically all have different difficulties. So we are not awarding more difficult items with higher scores. Then why should it be the case for partial credit items? For the Rasch model, multiple-choice items all receive a score of 1 as we assume that the items are equally discriminating (parallel ICCs). That is the rationale for awarding equal scores for the items, not the difficulty level. When we have partial credit items, there is an extra degree of freedom for us to set a maximum score (we don't have this freedom for multiple-choice items since every item has a score of 0 or 1). The decision then will need to be based on item discrimination (loosely speaking, how "good" an item is in separating good and poor students).

As for the number of "naturally" graded categories, the problem is that there could be many categories for one item and few categories for another item, leading to an unbalanced set of weights of the items in the test/questionnaire. As an example, if we are constructing a scale for measuring socio-economic status (SES) of households, geo-location may be divided into "urban", "rural", "remote", three categories, while number of people living in the home could be from 1 to, say, up to 10. House occupancy could be divided into "owning the home", "renting the home" and "other". If maximum score depends on the response categories, then some items will have more weight than other items but these weights will not reflect the relative importance of the items in measuring a construct.

So how does one decide on the maximum score (or weight) of an item? While item writers can typically gauge the difficulty of an item, it is often hard to estimate the discrimination of an item. The following are some guidelines.

- How well is the item related to the construct being measured? The more relevant an item is to the construct, the higher the maximum score should be.

- How much "information" can an item provide about students' ability/latent trait? It is often the case that open-ended items provide richer information about a student's capabilities than multiple-choice items do. That information can be used to separate students into several ability groups. If students' item responses can be clearly categorised into several increasing ability groups, then more score points can be awarded.

It should be noted that once the maximum score of a partial credit item is decided, score categories within an item must reflect increasing ability levels of students, and thus reflect increasing difficulties of the task. This may be a point that causes confusion: that the maximum score of an item relates to discrimination, but score categories within an item relate to difficulty.

Deciding on the maximum score of a partial credit item still involves a great deal of guesswork. As for dichotomous items, item statistics need to be checked to ensure the item responses fit the IRT model. Inappropriate maximum scores of partial credit items will be reflected in poor item fit statistics.

An Example Item Analysis of Partial Credit Items

The data set for this example came from a mathematics problem solving test for grade 5 students. The test had 48 questions, arranged in 3 rotated test booklets. In total, 1086 students took part in the test, but each item had around 500 student responses. The test had a mix of dichotomously and polytomously scored items. IRT and CTT analyses were conducted on the data set.

The following shows one example item ("Average") from the test and corresponding initial proposed scoring scheme:

Item 4: Item "Average"
Megan obtained an average mark of 81 for her four science tests. The following shows her scores for Tests 1, 3 and 4? What was her test score for Test 2? Show how you found your answer.

Test 1	Test 2	Test 3	Test 4	Average mark of 4 tests
84	?	89	93	81

As students were requested to provide their working in solving the item, students' responses contained a variety of approaches and answers. These responses

Table 9.1 Scoring scheme for item "Average"

Response	Proposed score
Correct analytic method and correct answer of 58	4
Trial-and-error method, but still obtained the correct answer	3
Correct analytic method, but computation error, resulting in incorrect answer	2
Computed the average of the three scores, but unable to proceed to produce the correct answer	1
Other responses	0

Table 9.2 Item statistics for the item "Average"

Score category	Frequency	Percentage	Pt biserial correlation	Average ability
0	183	0.33	−0.60	−0.78
1	108	0.19	−0.12	−0.22
2	36	0.06	0.12	0.43
3	23	0.04	0.09	0.40
4	209	0.37	0.57	0.67

were categorised according to test writers' views on the quality of the responses. In summary, an initial scoring guide was developed as shown in Table 9.1.

Item statistics for this item are shown in Table 9.2.

A few observations can be made from the item statistics in Table 9.2. First, very few students used the correct method but made a computational error leading to an incorrect answer (score category 2). Similarly, few students used the trial and error method and obtained the correct answer (score category 3). Second, the point biserial correlations for categories 2 and 3 are very similar. Third, the average abilities of students in categories 2 and 3 are very similar. These observations suggest that categories 2 and 3 can possibly be combined.

The fit statistics for this item are given in Table 9.3. The fact that the fit mean squares statistics are larger than 1 indicates that the item is not as discriminating as the model expects for an item with a maximum score of 4. This is further confirmed by the expected scores curve, as shown in Fig. 9.11. For high ability students, the observed score is lower than the expected score.

Based on these item statistics, a recoding of the score categories is made by collapsing categories 2 and 3 into a new category 2, and recoding the current category 4 as category 3. That is, the item has a maximum score of 3 after recoding.

Table 9.3 Fit statistics for item "Average"

Parameter	Infit mean squares	Infit t
δ_1	0.99	−0.14
δ_2	0.96	−0.72
δ_3	1.18	2.85
δ_4	1.14	2.34

Table 9.4 Item statistics for the item "Average", after recoding

Score Category	Frequency	Percentage	Pt biserial correlation	Average ability
0	183	0.33	−0.59	−0.78
1	108	0.19	−0.11	−0.20
2	59	0.11	0.17	0.47
3	209	0.37	0.54	0.66

Table 9.5 Fit statistics for item "Average", after recoding

Parameter	Infit mean squares	Infit t
δ_1	0.94	−0.76
δ_2	0.93	−1.55
δ_3	0.97	−0.52
δ_4	1.05	1.08

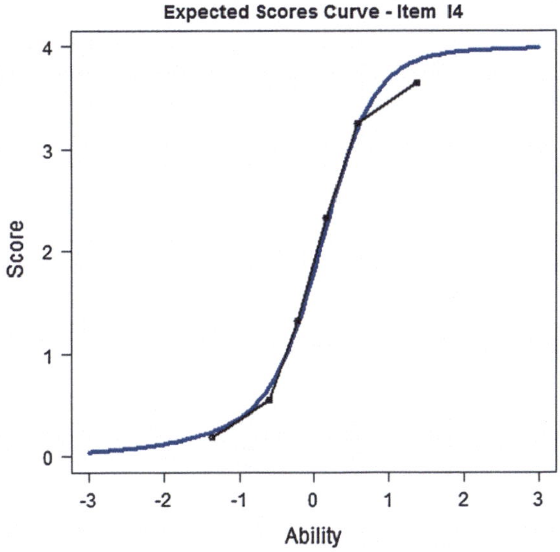

Fig. 9.11 Expected scores curve for item "Average"

The item statistics of the recoded item are shown in Table 9.4. It can be seen that both the point-biserial correlations and the average abilities show a nice progression with increasing scores.

The fit statistics after recoding are shown in Table 9.5. The fit statistics have improved after recoding, as the fit mean squares are closer to 1 and fit t values closer to 0 than before the recoding.

The expected scores curve after recoding is shown in Fig. 9.12, where the observed curve matches the expected reasonably well particularly for high ability students.

The analyses presented above show that the maximum score assigned to a partial credit item needs to be checked during the item analysis process. Frequently, recoding of partial credit categories is needed. Further, the perceived progression of response quality may not actually reflect increasing ability levels of students. In this example, students who used the correct method but obtained an incorrect answer through computational errors are of similar ability as students who obtained the correct answer using a trial-and-error approach. If we simply score the responses on the basis of correct/incorrect answer, we could have under-estimated the ability of students who made computational slips. These are all issues to be considered regarding scoring of item responses.

As an exercise, if we score the item simply based on correct and incorrect answers (i.e., categories 3 and 4 are recoded to 1; categories 1 and 2 are recoded to 0), we observe a gross overfit of the item. Figure 9.13 shows the expected scores curve. In this case, the maximum score for this item is 1. The observed curve is much more discriminating (steeper) than the expected curve, indicating that the maximum score can be increased. The fit statistics also shows a fit mean squares value less than 1, and fit t statistics large negative (see Table 9.6), indicating that the item is more discriminating than that expected of an item with a score of 1 (Fig. 9.13).

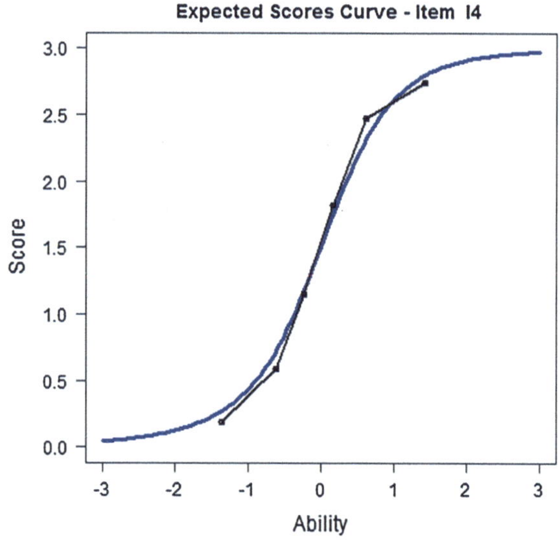

Fig. 9.12 Expected scores curve of item "Average", after recoding

Table 9.6 Fit statistics for item "Average", with dichotomous "correct/incorrect" scoring

Parameter	Infit mean squares	Infit t
δ_1	−0.88	−4.12

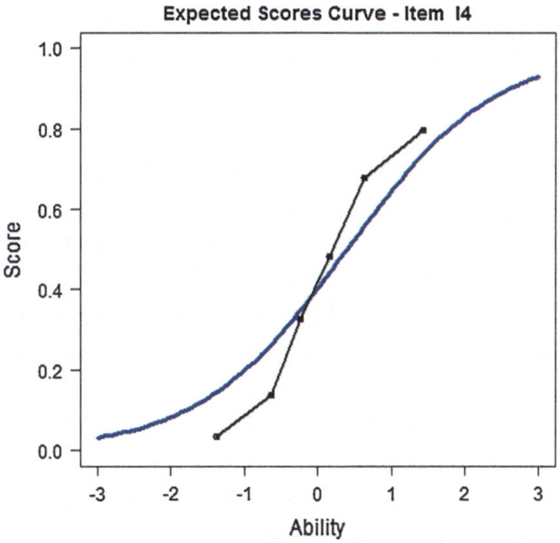

Fig. 9.13 Expected scores curve of item "Average", with "correct/incorrect" scoring

The example item presented here demonstrates that when the item responses provide more "information" about students' ability levels, partial credit scoring can be applied. If we only captured students' final answer, there is limited scope of dividing the item responses into ability groups.

On the other hand, we should stress that if a score category is found to be not "attractive" (i.e., few students in the category), this observation alone does not warrant the collapsing of the categories. If the item fits the IRT model, then collapsing categories will lead to a mis-fit of the item. If there is evidence to show that the item is not as discriminating as expected, then categories should be collapsed. In general, the relative frequencies of responses in score categories are unrelated to the decision of collapsing categories.

Rating Scale Model

In the partial credit model where the item parameters are expressed as $\delta_\bullet$ and τ parameters, as shown in Eq. (9.14),

$$\Pr(X_{ni} = x) = \frac{\exp \sum_{k=0}^{x} (\theta_n - (\delta_\bullet - \tau_{ik}))}{\sum_{h=0}^{m_i} \exp \sum_{k=0}^{h} (\theta_n - (\delta_\bullet - \tau_{ik}))} \quad (9.14)$$

τ_{ik} are often known as "step parameters". For the partial credit model, the step parameters are different across different items. This is the reason for the subscript i

and k for τ where i indicates the item number and k indicates the category number. A special case of the partial credit model is to constrain τ_{ik} to be the same across items. That is, we drop the subscript i so there is only one set of step parameters τ_k for all items. In this case the step "structure" for all items is the same. For example, we may want the "distance" between "strongly agree" and "agree" to have the same meaning across all items. We therefore use only one set of step parameters, τ_k, for all items. Such a model is commonly known as a rating scale model (Andrich 1978; Andersen 1997). While the rating scale model may have some desirable theoretical properties for some applications, in real-life, few data sets fit the model, as the model is much restricted as compared to the partial credit model.

Graded Response Model

A well-known IRT model for fitting data with ordered categories is Samejima's graded response model (GRM) (Samejima 1969, 1997). In contrast to the derivation of the PCM using probabilities for adjacent response categories, the graded response model is derived using cumulative probabilities for successfully completing up to step k of an item. If the cumulative probability functions are denoted by $P_k^*(\theta)$, then the probability of being in category k for a student with ability θ is given by

$$P_k(\theta) = P_k^*(\theta) - P_{k+1}^*(\theta) \tag{9.15}$$

The cumulative probability function can take the form of the normal ogive function or the logistic function. The graded response models typically have discrimination parameters as well as item difficulty parameters (category thresholds). Note that under GRM, the item category thresholds are never disordered.

Generalized Partial Credit Model

A generalization of the partial credit model to include discrimination parameters is the Generalized Partial Credit Model (GPCM). See Muraki (1992, 1997) for a description of this model. This model will be briefly discussed in Chap. 10.

Summary

This chapter explains the partial credit model from both theoretical and practical viewpoints. For the dichotomous model, the notion of item difficulty is clearly defined as the ability at which there is a 50% chance of obtaining the correct answer (or a score of 1). For the partial credit model, the notion of item difficulty becomes complex since there are several score points within an item. It may be easy to obtain a lower score for the item, but difficult to obtain a high score for the item so the concept of an overall difficulty of an item needs to be clarified. The δ parameters have been used as difficulty parameters in some situations, but these parameters pose problems for interpretation when middle response categories have few respondents, leading to the reversal of the δ parameters (so-called dis-ordered thresholds). To avoid the problems of the dis-ordering of δ, the Thurstonian thresholds are sometimes preferred as difficulty measures for each score category.

For scoring a partial credit item, it should be noted that the maximum score assigned for an item is the weight of the item in relation to other items in the test. The maximum score should not be arbitrarily determined. Instead, the maximum score of an item should relate to the discriminating power of an item, and not the difficulty of an item. Various checks should be made to ensure that the maximum score assigned results in adequate fit of the item to the model. In addition, to determine the score categories within an item, one should ensure that increasing item difficulty (hence increasing ability) matches with increasing score categories. This can be checked by examining the point-biserial correlations and average abilities for the score categories within the item. Such item statistics are typically provided through a classical test theory (CTT) analysis and IRT item analysis. Note that there is only one way of scoring an item that will fit the PCM model, so we cannot simply assign category scores in any way we would like.

The example provided in this chapter on evaluating the appropriateness of category scoring is somewhat complex, as many statistics need to be taken into account. This is the case since the Rasch model does not provide a discrimination index for each item. Information on the discriminating power of an item can only be obtained through plotting the expected scores curve, CTT and fit statistics. If an IRT model provides direct estimations of item discrimination indices, then we will not need to take the roundabout way of estimating the discrimination of an item. This is precisely what a two-parameter IRT model (2PL) does. In fact, when we use the partial credit model and have the task of assigning a weight (maximum score) to an item, we are already thinking in the framework of the two-parameter model. The next chapter explains the two-parameter model and contrasts it with the Rasch model.

Discussion Points

1. For a partial credit item, there is a requirement that score categories are "ordered". Discuss the meaning of "ordered" in this requirement. (You can consider the meaning of "ordered" in relation to ability, difficulty, expected score, probability of success, etc.)
2. A serious misconception about the treatment of dis-ordered thresholds is that when two categories have reversed δ, one must reverse-score the categories

(e.g., code category 2 as 1, and category 1 as 2). Discuss why this should NOT be done. Under what situations should the scoring be reversed?

Exercises

Q1. Plot the ICC of a 3-category PCM item in a spreadsheet with δ_1 and δ_2 as parameters. For example,

delta1	delta2
-1	1.3

ability	pr(0)	pr(1)	pr(2)
-3	0.879375	0.11901	0.001615
-2.9	0.868198	0.129855	0.001947
-2.8	0.856136	0.141518	0.002345
-2.7	0.843149	0.154029	0.002821
-2.6	0.829199	0.167412	0.003389
-2.5	0.814252	0.181684	0.004064
-2.4	0.79828	0.196853	0.004867
-2.3	0.781263	0.212919	0.005818
-2.2	0.76319	0.229868	0.006941
-2.1	0.744059	0.247676	0.008266
-2	0.723878	0.2663	0.009822
-1.9	0.70267	0.285684	0.011645
-1.8	0.680471	0.305755	0.013774
-1.7	0.657329	0.32642	0.016251

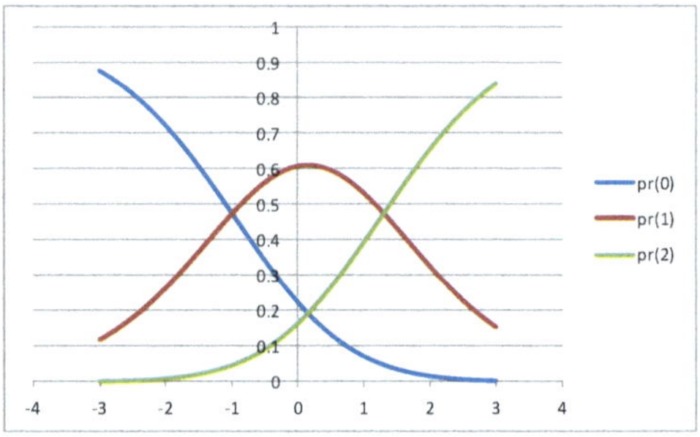

Change the values of δ_1 and δ_2 and see how the ICCs change. In particular, try ordered δ_1 and δ_2, and dis-ordered δ_1 and δ_2.

An extension of this exercise is to add simulated item responses for each student, and plot the observed item characteristic curves as well.

Q2. Indicate whether you agree or disagree with each of the following statements

A response category has very few students. So the response category should be collapsed with an adjacent category	Agree/disagree
Collapsing score categories of an item should have no impact on the fit of the item	Agree/disagree
Item A has a maximum score of 4. Item B has a maximum score of 2. Item A must be more difficult than Item B	Agree/disagree
When dis-ordered thresholds are observed, this indicates that the item does not fit the PCM	Agree/disagree
When dis-ordered thresholds are observed, this indicates that high ability students have a lower expected score than low ability students	Agree/disagree
When dis-ordered thresholds are observed, we should reverse score the response categories	Agree/disagree

References

Adams RJ, Wu ML, Wilson M (2012) The Rasch rating model and the disordered threshold controversy. Educ Psychol Measur 72:547–573

Andersen EB (1997) The rating scale model. In: van der Linden WJ, Hambleton RK (eds) Handbook of modern item response theory. Springer, New York, pp 67–84

Andrich DA (1978) A rating formulation for ordered response categories. Psychometrika 43: 561–573

Masters GN (1982) A Rasch model for partial credit scoring. Psychometrika 47:149–174

Masters GN, Wright BD (1997) The partial credit model. In: van der Linden WJ, Hambleton RK (eds) Handbook of modern item response theory. Springer, New York, pp 101–121

Muraki E (1992) A generalized partial credit model: application of an EM algorithm. Appl Psychol Meas 16:159–176

Muraki E (1997) A generalized partial credit model. In: van der Linden WJ, Hambleton RK (eds) Handbook of modern item response theory. Springer, New York, pp 153–164

Samejima F (1969) Estimation of latent ability using a response pattern of graded scores. Psychometric Monogr, No, p 17

Samejima F (1997) Graded response model. In: van der Linden WJ, Hambleton RK (eds) Handbook of modern item response theory. Springer, New York, pp 85–100

Verhelst ND, Glas CAW, de Vries HH (1997) A steps model to analyse partial credit. In: van der Linden WJ, Hambleton RK (eds) Handbook of modern item response theory. Springer, NY, pp 123–138

Verhelst ND, Verstralen HHFM (1997) Modeling sums of binary responses by the partial credit model. Cito Measurement and Research Department Reports, pp 97–107. Cito

Further Reading

Ostini R, Nering ML (2006) Polytomous item response theory models. Sage Publications

Chapter 10
Two-Parameter IRT Models

Introduction

The Rasch model is sometimes also called the one-parameter IRT model in that the probability of success as a function of the ability θ has only one parameter (the item difficulty parameter) estimated for each item, as shown in Eq. (10.1)

$$p = P(X = 1) = \frac{\exp(\theta - \delta)}{1 + \exp(\theta - \delta)} \tag{10.1}$$

The Rasch model assumes that items all have the same discrimination, in that the item characteristic curves are parallel, as shown in Chap. 7. In contrast, a discrimination parameter can be incorporated into Eq. (10.1) thereby extending it to a more general mathematical model as seen in Eq. (10.2)

$$p = P(X = 1) = \frac{\exp(a(\theta - \delta))}{1 + \exp(a(\theta - \delta))} \tag{10.2}$$

In Eq. (10.2), the parameter a is called the discrimination parameter (or slope parameter), in addition to the item difficulty parameter δ. This model is usually known as the 2PL model (2-parameter logistic model). The parameter a is a scale factor of the ability scale, since it is multiplied to $(\theta - \delta)$, where θ and δ are in logit unit on the ability scale. Such a multiplying factor has the effect of stretching or shrinking the ability scale, in the same way as one imagines using the Windows re-size tool (⇔) to change the horizontal scale of a picture, as illustrated in Fig. 10.1. Two items with the same item difficulty ($\delta = 0.8$) but different discrimination parameters, $a = 0.6$ and $a = 1.7$ respectively are shown in the left and right graphs of Fig. 10.1.

It can be seen from Fig. 10.1 that the larger the value of a, the steeper the ICC, and the more discriminating an item is. In contrast, a very flat ICC indicates that

M. Wu et al., *Educational Measurement for Applied Researchers*,
DOI 10.1007/978-981-10-3302-5_10

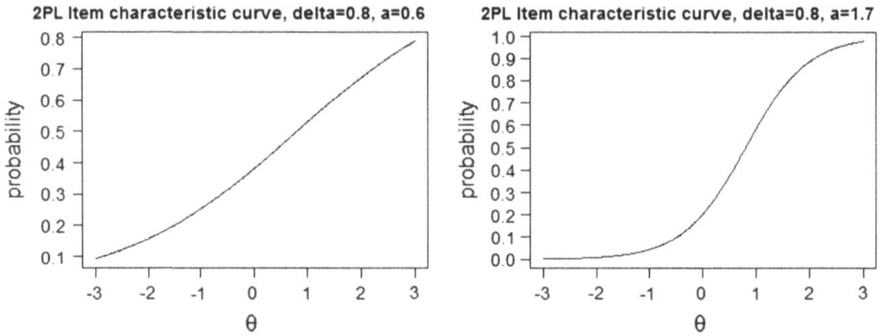

Fig. 10.1 2PL ICC with a = 0.6 (*left graph*) and a = 1.7 (*right graph*)

low and high ability students have similar chances of obtaining the correct answer, so the item is not very discriminating. To make the left-side curve in Fig. 10.1 steeper, we need to shrink the scale. To make the right-side curve flatter, the graph needs to be stretched horizontally. In this way, it can be seen that highly discriminating items can separate students more than low discrimination items can.

Discrimination Parameter as Score of an Item

In Chap. 9, partial credit item scoring is discussed. In particular, the maximum score of a partial credit item is a weight of the item in the whole test, and this maximum score should be set in relation to the item discrimination (not item difficulty). Recall that in Chap. 9, the probability of scoring a 2 in a 3-category (0, 1, 2) item is

$$p_2 = \Pr(X = 2) = \frac{\exp(2\theta - (\delta_1 + \delta_2))}{1 + \exp(\theta - \delta_1) + \exp(2\theta - (\delta_1 + \delta_2))} \tag{10.3}$$

In Eq. (10.3), it can be seen that the maximum score of a partial credit item (in this case, 2) relates to the multiplier of θ (2θ in the numerator of Eq. (10.3)), similar to the a parameter in the 2PL model. In this sense, the a parameter of 2PL model can be regarded as the score of an item. One difference between the 2PL model and the partial credit model is that item scores are estimated from the item response data in 2PL, and not set by the test writer as for the partial credit model.

More generally, item "scores" or item weights in 2PL are estimated for every item, including dichotomous and partial credit items. The following is an example of the differences between the Rasch model and the 2PL model for a set of dichotomously scored items.

An Example Analysis of Dichotomous Items Using Rasch and 2PL Models

The data set of this example contains item responses of 2987 students to a mathematics test with 13 multiple-choice items. First, a Rasch analysis is carried out. In this analysis, each item is scored 1 for correct answer and 0 for incorrect answer. Table 10.1 shows the item statistics for the 13 items from the Rasch and CTT analyses.

It can be seen from Table 10.1 that at least a few items do not fit the Rasch model well. For example, item 2 and item 13 have large fit mean squares values and lower discrimination indices. In contrast, items 4, 5, and 6 "over-fit" the model, with high discrimination indices. Figure 10.2 shows the ICC of item 2 and item 5 as two example items.

Figure 10.2 shows that an "under-fit" item corresponds to low discrimination, or flatter observed ICC, and an "over-fit" item corresponds to high discrimination, or a steep observed ICC. As discussed in Chap. 8, the residual-based fit statistics reflect the slope of the observed ICC against the theoretical ICC.

Item 2 and item 5 are shown in Fig. 10.3 to offer some suggestions as to why one item is not very discriminating while the other one is.

The first observation is that item 2 is a much more difficulty item than item 5 (see item difficulty estimates in Table 10.1). Item 2 is a word problem, requiring students to know the number of days in June and July. Further, it is unclear whether the two end days should both be counted. That is, if a person arrives on the 1st of June and leaves on the 2nd, it is unclear whether this is counted as one day or two days. In this test, the correct answer is 53 days, which includes both end days. In contrast, item 5 is a computation item. It requires students to know subtraction procedures without borrowing. There is a clear correct answer. It is not a difficulty item, but yet this item is more discriminating than item 2. That is, the computation

Table 10.1 Rasch and CTT item statistics of a set of dichotomous items

	Difficulty	Infit MS	Infit t	Pt-bis corr
M_1	−1.43	1.03	1.24	0.45
M_2	0.80	1.30	13.66	0.32
M_3	−0.44	0.96	−2.37	0.58
M_4	−0.02	0.90	−5.66	0.62
M_5	−0.73	0.88	−6.58	0.62
M_6	0.28	0.92	−4.34	0.61
M_7	−0.28	0.99	−0.31	0.55
M_8	0.13	0.97	−1.70	0.57
M_9	0.46	0.99	−0.49	0.56
M_10	−0.18	0.94	−3.61	0.59
M_11	0.19	1.00	−0.16	0.56
M_12	0.53	0.96	−1.97	0.57
M_13	0.67	1.20	9.72	0.38

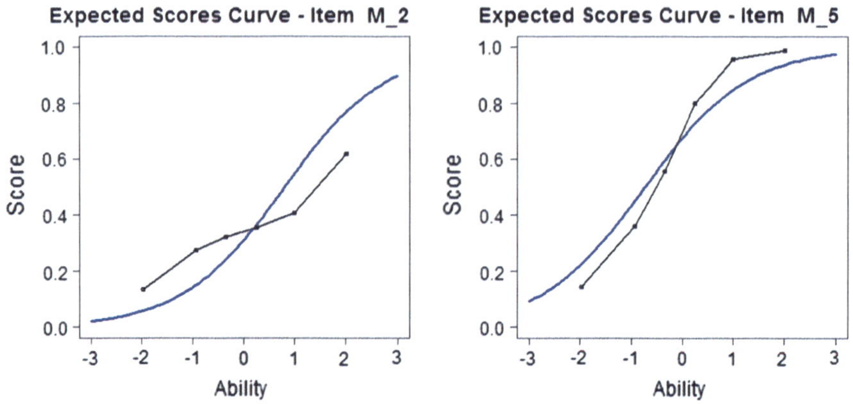

Fig. 10.2 ICC of two example items showing "under-fit" and "over-fit"

Item 2	Item 5
Mike reached Sydney on 13th June in the morning and left on 4th August in the night. For how many days did Mike stay in Sydney?	Solve the following 7895 - 5704 ______
(1) 53 days (2) 52 days (3) 51 days (4) 50 days	(1) 1191 (2) 2191 (3) 2101 (4) 1101

Fig. 10.3 Item 2 and item 5 in the example test

item can separate low and high ability students better than the "number of days in the calendar" item. One may also conjecture that knowing the number of days in June and July may not be directly related to a student's general mathematics ability.

Such a post hoc analysis of items can suggest reasons for item difficulty and item discrimination. However, prior to the administration of test items, it may be difficult to guesstimate item discrimination. Test writers can often gauge the item difficulty from an analysis of cognitive load for an item or from curriculum progression of an item topic. But item discrimination is not so easy to predict. In general, constructed response items have higher discrimination than multiple-choice items (irrespective of item difficulty) because of the chance of guessing in multiple-choice items.

We further note that item difficulty parameter and item slope parameter are two different and unrelated statistics. Unlike CTT item discrimination/point-biserial correlation statistics which relate to item difficulty, the IRT δ and a parameters are "unrelated" in the sense that an easy or difficult item can have high or low values of a. Under IRT, the notions of item difficulty and item discrimination are two distinct concepts. We discuss this distinction further in the latter part of this chapter.

2PL Analysis

A 2PL analysis is carried out on the same data set. Table 10.2 shows estimated item difficulties, slope parameters and fit indices.

Figure 10.4 shows the ICCs of item 2 and item 5 using the 2PL model to fit the item responses.

Comparing 2PL model with the Rasch model, a number of observations can be made. First, the item characteristic curves in Fig. 10.4 show that the theoretical (or modelled) expected scores curves can have different slopes across different items. As a result, the theoretical curves from a 2PL analysis fit the observed curves better than for the Rasch model. Checking the fit statistics in Table 10.2, it can be seen that all items show good fit. In fact, since the residual-based fit statistics detect departure of the slope of the observed ICC from the expected ICC, these fit statistics will necessarily show good fit when the theoretical ICC can have varying slopes to match the observed ICC. Consequently, residual-based fit statistics are not useful for checking item fit for 2PL models.

To further demonstrate the relationship between residual-based fit statistics and item discrimination, the Rasch infit mean squares are plotted against the 2PL slope parameters across all items. Figure 10.5 shows this plot.

Figure 10.5 shows that the larger the residual-based fit statistic, the lower the 2PL slope parameter. In other words, when an item "under-fits" the Rasch model, the item is not as discriminating as the Rasch model expects, the 2PL model assigns

Table 10.2 2PL item statistics of a set of dichotomous items

	Difficulty	Slope parameter	Infit MS	Infit t
M_1	−1.33	1.05	0.99	−0.28
M_2	0.64	0.43	1.00	−0.09
M_3	−0.50	1.61	1.02	0.91
M_4	−0.04	1.94	1.00	−0.11
M_5	−0.95	2.07	0.99	−0.23
M_6	0.32	1.74	1.00	−0.11
M_7	−0.28	1.28	1.00	0.24
M_8	0.13	1.44	0.98	−1.02
M_9	0.47	1.34	1.01	0.50
M_10	−0.20	1.51	1.00	0.16
M_11	0.19	1.33	1.01	0.42
M_12	0.56	1.49	1.00	−0.19
M_13	0.55	0.58	1.00	0.17

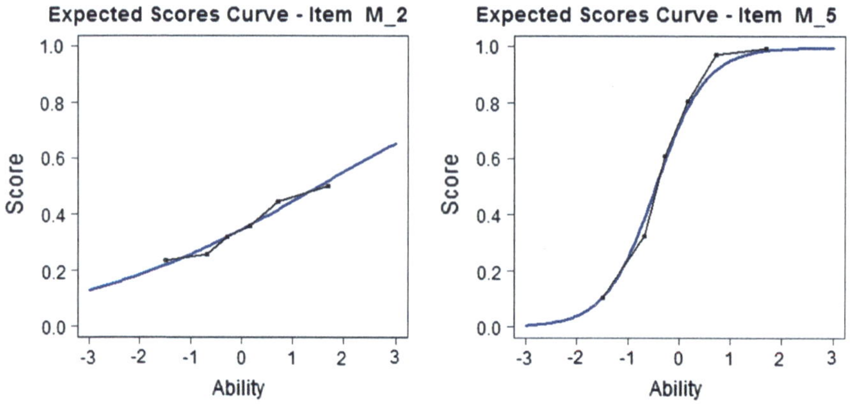

Fig. 10.4 ICC of item 2 and item 5 using 2PL model

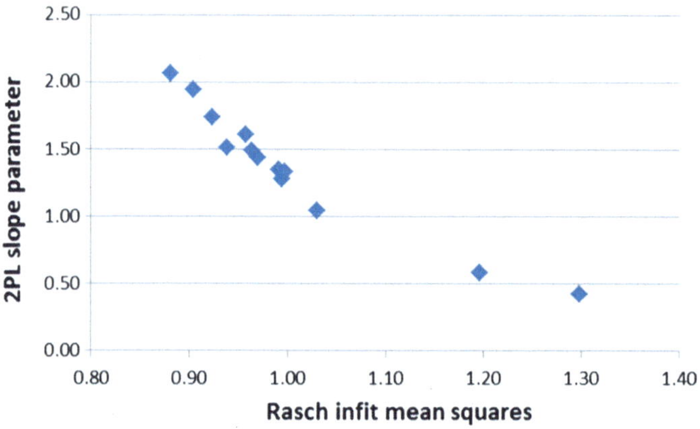

Fig. 10.5 Rasch fit statistics plotted against 2PL slope parameter

a lower weight (score) for the item. Essentially, what the 2PL model does is to estimate weights for the items according to the discriminating power of the items. The "worse" (i.e., little discriminating power) an item is, the smaller the item weight. In the case of the example, item 5 has a much large weight (score) than item 2 (see Table 10.2). This makes a great deal of sense. If we believe that item 2 has ambiguous correct answer, lowering the weight of this item is one way to provide "fairer" scoring. In summary, while each item under the Rasch model contributes equally towards the total test score, items under the 2PL models contribute differently according to the discriminating power of the items.

A plot of the CTT discrimination index against the 2PL slope parameter for each item shows again that the slope parameter is a measure of item discrimination. See Fig. 10.6.

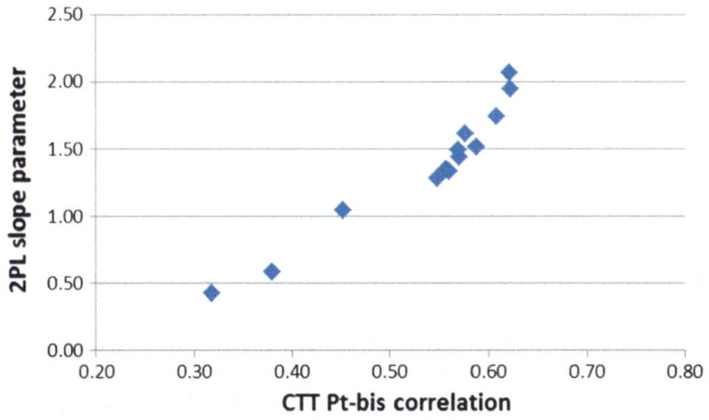

Fig. 10.6 CTT point-biserial correlation plotted against 2PL slope parameter

To further clarify the relationship between the Rasch model and the 2PL model, two more plots are shown. The first is a plot of the item difficulty estimates obtained from the Rasch model and the 2PL model. See Fig. 10.7.

Figure 10.7 shows that the item difficulty parameters obtained from the Rasch model and the 2PL model correlate well.

Figure 10.8 shows a plot of item difficulty against slope parameter under the 2PL model.

Figure 10.8 shows that there is no discernible relationship between item difficulty and the discrimination parameter. This observation reinforces the recommendation in Chap. 9 that the maximum score of an item should not be dependent on the item difficulty. In fact, Fig. 10.8 shows that for two difficult items, their slope parameters are low and hence their weights (scores) are lowered.

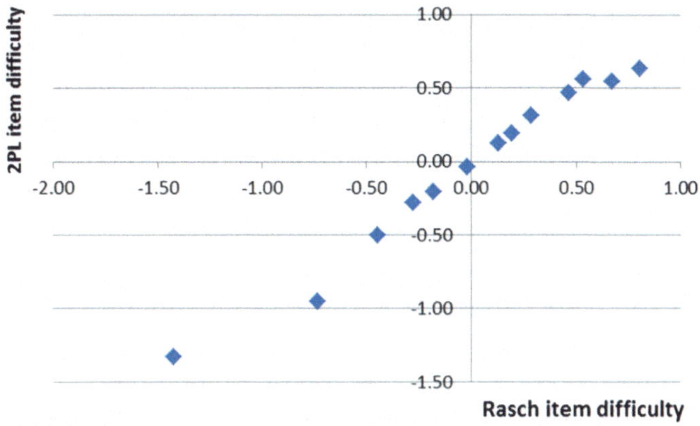

Fig. 10.7 Rasch item difficulty plotted against 2PL item difficulty

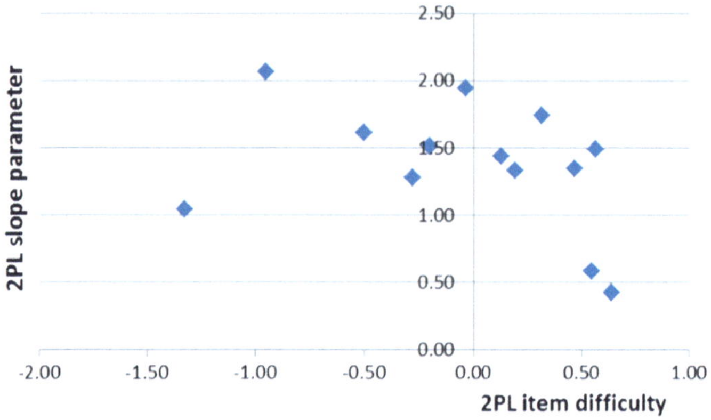

Fig. 10.8 2PL item difficulty against 2PL slope parameter

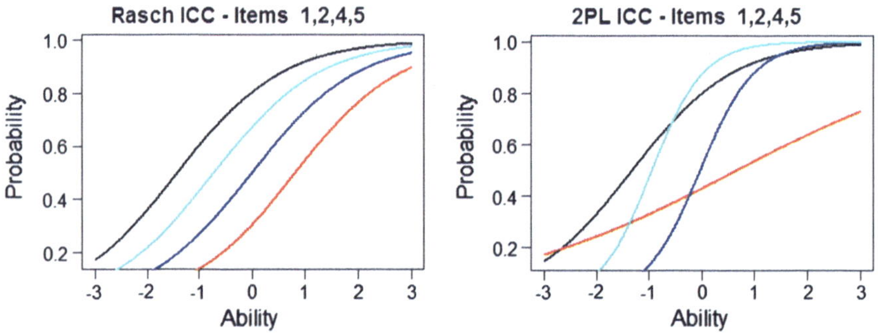

Fig. 10.9 Overlay of theoretical ICCs for the Rasch model (*left graph*) and the 2PL model (*right graph*)

Finally, to illustrate the difference between the Rasch model and the 2PL model, a set of theoretical ICCs are plotted in the same graph (i.e., overlay) to show the parallel and non-parallel nature of the curves for the Rasch and 2PL models respectively.

Figure 10.9 shows that the Rasch model fits parallel theoretical ICCs irrespective of the slopes of the observed ICCs, while the 2PL model fits theoretical ICCs with different slopes to match that of the observed data.

A Note on the Constraints of Estimated Parameters

In the discussion of the Rasch model in Chap. 7, the indeterminacies of the location and scale of the latent trait measures are explained. For the 2PL model, similar issues of indeterminacies apply. In particular, the scale factor of the latent trait needs some clarification. As discussed in Chap. 7, for the Rasch model, the discrimination parameter, a, is commonly set to 1 (or 1.7, see Chap. 7 hands-on Practices Task 2). For the 2PL model, a is estimated for each item so there are different values of a for different items. The constraint for the scale can be set in a number of different ways. For example, the average of the set of a parameters can be fixed. Alternatively, the scale can be set by fixing the variance of the ability distribution to an arbitrary value, typically 1, in much the same ways as setting the mean of the ability distribution to 0 for the estimation of the δ parameters. Different software programs will have different ways of setting the scale constraint. So you need to check the manuals of specific programs for details. We have found that fixing the variance of the ability distribution leads to better convergence in the parameter estimation process, although fixing the variance can make comparisons of item scores more difficult. In any case, the slope parameters can always be transformed again after estimation. The following is an example.

In Table 10.2, the slope parameters are estimated with the scale constraint of setting the variance of the ability distribution to 1. The average of the slope parameters for the 13 items is 1.37. If we want to compare the slope parameters (i.e., item scores) to the relative Rasch item scores of 1, we can transform the slope parameters to have an average of 1. In Table 10.3, we divide each slope parameter by the average of the slope parameters (1.37) so that the transformed slope parameters have an average of 1. The transformed item scores are particularly useful for partial credit items, as explained in latter sections of this chapter. Note that such transformation of the scale parameters is not necessary for model fitting unless comparisons of the slope parameters across different scaling runs are made.

Consequently, the slope parameters from different scaling runs may not be directly comparable because of the indeterminacy of the scale unit unless

Table 10.3 Transformed slope parameters with an average of 1

Item	Slope parameter	Transformed slope parameter
M_1	1.05	0.77
M_2	0.43	0.31
M_3	1.61	1.18
M_4	1.94	1.42
M_5	2.07	1.51
M_6	1.74	1.27
M_7	1.28	0.93
M_8	1.44	1.05
M_9	1.34	0.98
M_10	1.51	1.10
M_11	1.33	0.97
M_12	1.49	1.09
M_13	0.58	0.42
Average	1.37	1.00

transformations are carried out. To get an idea of the overall discriminating power of a test instrument, the test reliability index is a better indicator.

A Note on the Parameterisation of Item Difficulty Parameters Under 2PL Model

In Eq. (10.2), the numerator, $\exp(a(\theta - \delta))$, expresses the slope parameter as a multiplier of $(\theta - \delta)$. The argument of the exponential function can also be expanded as $(a\theta - a\delta)$. In some software packages, the item difficulty parameter reported for 2PL is $a\delta$ rather than δ. Check software documentations for the parameterisation, as different parameterisations can lead to different interpretations of the item difficulty parameters.

Impact of Different Item Weights on Ability Estimates

Under the Rasch model, raw scores on a test are "sufficient statistics" for ability estimates (see Chap. 7). That is, students with the same raw score on a test will have the same ability estimate, irrespective of which items they answered correctly, since all items have the same weight in the test. Under the 2PL model, students with the same raw score may not necessarily have the same ability estimate; it will depend on the particular set of items a student answered correctly. If the items answered correctly have more weights, then the ability will be higher. This makes sense in that if an item does not discriminate students (say, the responses are random guesses for that item), then obtaining the correct answer on this item does not indicate a more able student. So this item "counts" less towards the ability estimate. Table 10. 4 shows weighted likelihood ability estimates (WLE) for selected students from the example data set.

The ability estimates from 2PL models are likely to be closer to students' "true" abilities than Rasch ability estimates are, since the estimation takes into account the amount of "information" provided by each item. However, providing different ability estimates for the same raw score may pose a problem for examination officials who may need to explain to the layperson how ability estimates are derived. In providing such explanations, it is inevitable to acknowledge that items are of varying "quality" in the test. For high-stake examinations, this issue needs to be considered.

Table 10.4 Ability estimates for selected students with a raw score of 10 out of 13

Student id	Rasch WLE ability	2PL WLE ability
15	1.20	0.88
17	1.20	1.16
18	1.20	1.26

Choosing Between the Rasch Model and 2PL Model

In Chaps. 6 and 7, the desirable measurement properties of the Rasch model have been presented. So why would one choose the 2PL model over the Rasch model? Here are some possible reasons. When the Rasch model is used, the good properties of the model can only be realised if the data fit the model. Given a particular data set, such as the example provided in this chapter, mis-fitting items do not get "fixed" by running a Rasch analysis. That is, the properties of the Rasch model are not attained since the observed ICCs are not "parallel" even though the theoretical ICCs are (forced to be). Under such circumstances, one may decide to choose a model that fit the data better, and make the best use of the information available, since choosing a mis-fitting model will not provide the properties of the model. Rasch models are useful for the construction of an instrument if there are possibilities of modifying and deleting items. If a test has already been administered and the data do not fit the Rasch model, there is no gain in using the Rasch model. In fact, there are some gains in using a model that fit the data.

In real-life, no item response data will fit a theoretical model perfectly, since the models are mathematical functions. The more parameters a model has, the more likely the data will fit the model. There is no real-data set that will fit the Rasch model perfectly (nor a 2PL model, for that matter), so we need to make an assessment of how good a fit is good enough. From a practical point of view, it probably makes little difference whether Rasch model or 2PL models are fitted if we have quality items in a test. If the data fit the Rasch model well, then a 2PL model fitted to the data set will also have similar slopes across items. Although from a theoretical point of view, the good properties of measurement should be upheld at least as a goal to achieve when instruments are constructed. In practice, there needs not be a clear demarcation when it comes to choosing IRT models, but a good understanding of the implications of each model and model-fit is important.

2PL Models for Partial Credit Items

An extension of the 2PL to the partial credit items is the generalised partial credit model (GPCM) (Muraki 1992). As for the dichotomous case, a discrimination parameter is added to the partial credit model presented in Chap. 9. Eq. (10.3) shows the GPCM.

$$\Pr(X_{ni} = x) = \frac{\exp \sum_{k=0}^{x} a_i(\theta_n - \delta_{ik})}{\sum_{h=0}^{m_i} \exp \sum_{k=0}^{h} a_i(\theta_n - \delta_{ik})} \quad (10.3)$$

Dropping the index i for item number for simplicity, a 3-category partial credit item has the following probabilities (see Eqs. (10.4)–(10.6)).

$$p_0 = \Pr(X=0) = \frac{1}{1 + \exp a(\theta - \delta_1) + \exp a(2\theta - (\delta_1 + \delta_2))} \tag{10.4}$$

$$p_1 = \Pr(X=1) = \frac{\exp a(\theta - \delta_1)}{1 + \exp a(\theta - \delta_1) + \exp a(2\theta - (\delta_1 + \delta_2))} \tag{10.5}$$

$$p_2 = \Pr(X=2) = \frac{\exp a(2\theta - (\delta_1 + \delta_2))}{1 + \exp a(\theta - \delta_1) + \exp a(2\theta - (\delta_1 + \delta_2))} \tag{10.6}$$

An Example Data Set

As an example, the data set analysed in Chap. 9 is re-run using the generalised partial credit model. Table 10.5 shows estimates of the slope parameters, a, and the item difficulty parameters for the first 10 items. The slope parameters have been transformed (scaled up) so the maximum score on the test is 68, matching that of the Rasch model.

Under the generalised partial credit model, a slope (or discrimination) parameter is estimated for every item, despite the number of response categories within the item. In the example give in Table 10.5, the value of the slope parameter varies across items in relation to the discriminating power of the item. As an example to illustrate the differences between the 2PL (GPCM) and the Rasch model (PCM), Fig. 10.10 shows a comparison of the ICCs between the two models for item 23 in the data set.

The slope parameter for item 23 is 0.49, indicating that the weight estimated for this item under GPCM is smaller than that assigned by the PCM. That is, the item is not as discriminating as the PCM model assumes. Similar to the 2PL dichotomous items, a weight is applied to weigh up or down the contribution of a PCM item to the total score. Within the item though, the weights of the categories are still in integer multiples, i.e., 1, 2, 3, etc. For example, for item 23, the weight (or score) of

Table 10.5 Slope parameter and item difficulty parameters

Item	a (transformed)	δ_1	δ_2	δ_3	δ_4
1	0.76	−1.58			
2	1.04	−1.93			
3	0.95	0.70	0.73		
4	0.70	0.24	1.03	0.58	−1.88
5	0.55	2.07	−2.76		
6	1.14	−1.56			
7	0.94	−0.74			
8	1.30	−1.88			
9	1.04	0.47			
10	0.86	0.28	0.42		

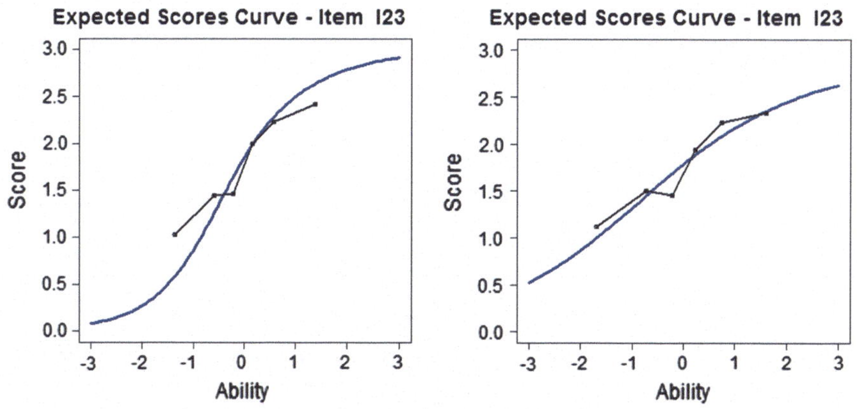

Fig. 10.10 Item I23 ICC–PCM model (*left graph*) and GPCM (*right graph*)

category 1 is 0.49 (the slope parameter), the weights of categories 2 and 3 are 2×0.49 and 3×0.49, respectively. That is, the coefficient of θ in Eqs. (10.3)–(10.6) is $ak\theta$, where a is the slope parameter and k is the item category number.

For this data set, the test reliability has increased a little from 0.82 for Rasch model (PCM) to 0.83 for 2PL (GPCM) model. Using 2PL will often increase the test reliability a little as more weights are assigned to more discriminating items.

A More Generalised Partial Credit Model

For the GPCM, one discrimination parameter is estimated for each item. That is, in Eq. (10.3), the slope parameter a_i has a subscript i for item i. However, if the subscript of the slope parameter is ik, then a_{ik} denotes a slope parameter for category k of item i.

$$\Pr(X_{ni} = x) = \frac{\exp \sum_{k=0}^{x} a_{ik}(\theta_n - \delta_{ik})}{\sum_{h=0}^{m_i} \exp \sum_{k=0}^{h} a_{ik}(\theta_n - \delta_{ik})} \tag{10.7}$$

In this case, there is a weight (or score) assigned to each score category of an item, and not just one weight for the whole item. This is a more general model than the GPCM, since different categories within an item can have different weights. Such models have been implemented in TAM (Kiefer et al. 2013) and in ConQuest (Wu et al. 2007) software programs. In this book, we will call this model KPCM for category level PCM. This model is similar to the idea of Bock's nominal response model (Bock 1972).

As an illustration, we use the data set in Chap. 9 and re-analyse using KPCM. The corresponding results for the first 10 items are shown in Table 10.6.

Table 10.6 Discrimination parameters at item category level under KPCM

Item i	a_{i1}	a_{i2}	a_{i3}	a_{i4}
1	0.76			
2	0.98			
3	0.58	2.18		
4	1.12	3.10	2.64	2.95
5	0.84	1.08		
6	1.13			
7	0.91			
8	1.24			
9	1.02			
10	0.93	1.71		

Table 10.6 shows the slope parameters at item category levels. As discussed, the parameters are regarded as weights, or scores, for the item categories. To put these into perspective, we compare the scores of item 4 under the PCM, GPCM and KPCM models. This item has 5 response categories, 0, 1, 2, 3, 4. Under the partial credit model, the scores are assigned and not estimated, so the scores for the five categories are 0, 1, 2, 3, 4. Under the generalised partial credit model, the scores are (see Table 10.5) 0, 0.70, 1.40 (0.70 × 2), 2.10 (0.70 × 3), 2.80 (0.70 × 4) for the five categories. Under the KPCM, the scores are estimated separately for the item categories, and these are (0), 1.12, 3.10, 2.64, 2.95.

In Chap. 9, it is shown that item 4 in the data set shows mis-fit under PCM and that the item is not as discriminating as the Rasch model expects using the assigned category scores. Consequently, a collapsing of categories and lowering of the maximum score lead to a better fit of the item. In this chapter, the estimated scores under GPCM and KPCM suggest that the maximum score should be lower. In particular, under KPCM, the scores for categories 2, 3 and 4 are similar. If the PCM is still the preferred model, then at least KPCM can suggest how the categories can be collapsed. In contrast, GPCM lowers the maximum score of the item, but keeps the *relative* weights at the category levels in integer multiples (i.e., 0, 1, 2, 3, 4).

Since KPCM has more parameters, the model necessarily will provide a better fit to the data. Figure 10.11 shows the expected scores curve for item 4 under KPCM.

Using KPCM, the reliability has increased slightly to 0.835.

A Note About Item Difficulty and Item Discrimination

Occasionally, there are confusions between the concepts of item difficulty and item discrimination. In particular, such confusion may arise when item-person maps are interpreted. Figure 10.12 shows an item-person map for two hypothetical partial credit items.

The item thresholds are plotted for two partial credit items where 1.1 refers to item 1, step 1, and 1.2 refers to item 1, step 2, etc. For this example, it does not

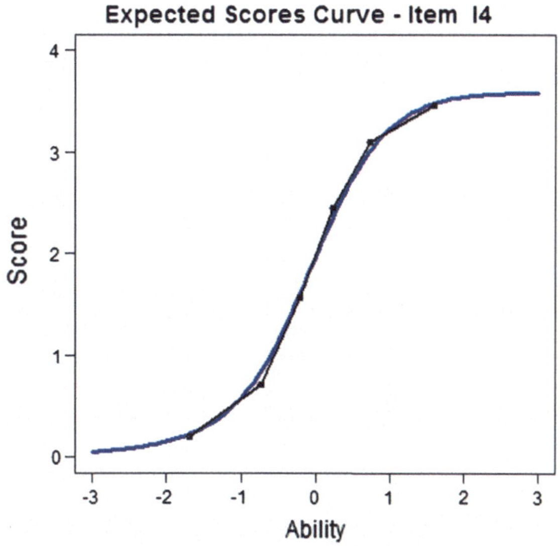

Fig. 10.11 Expected scores curve for item 4 under KPCM

matter whether we use the δ_{ik} parameters or γ_{ik} parameters for step difficulties of a partial credit item. Note that 1.1 and 1.2 are close together, and 2.1 and 2.2 are far apart. In this case, some may take this observation to infer that item 2 is more *discriminating* than item 1. This is not an appropriate interpretation. The locations of the item thresholds are item difficulty measures, just like for the dichotomous case. They do not provide any information about item discrimination. In the dichotomous case, the locations of items on the item-person map indicate item difficulty. In fact, if the items fit the Rasch model, then all items have the same discriminating power. However, the ICC of an item is the steepest at $\theta = \delta$. That is, an item has more discriminating power for the ability range close to the item difficulty of an item. But all items have the same overall discriminating power in a Rasch model. Similarly, for partial credit items, the locations of the thresholds can indicate the ability range at which the item is most discriminating. But the locations of the thresholds do not tell us anything about the overall discriminating power of an item. If we plot the expected scores curves for item 1 and item 2, the curve for item 1 appears to be steeper than that for item 2 (Fig. 10.13).

The expected scores curves in Fig. 10.13 shows that item 1 is steeper than item 2 at around the ability region of (−1, 0 1). Outside this region, the expected scores curve for item 2 tends to be steeper. If the two items fit the Rasch model, then, since the maximum score for both items is 2, the weights of these two items are equal, so the two items provide the same overall discriminating power. Moreover it can be observed from the figure that at different ability levels, the two items provide different discriminating power. Nevertheless, from Fig. 10.13, there is no evidence that item 2 is a better item just because the thresholds are further apart.

To sum up, the item-person map provides item difficulty information. It does not show how well an item can discriminate the respondents. To further clarify item

Item-person map

```
               XX|                                    |
                X|                                    |
               XX|                                    |
              XXX|                                    |
              XXX|                                    |
             XXXX|                                    |
             XXXX|                                    |
          XXXXXXX|                                    |
 1         XXXXXX|       2.2                          |
           XXXXXX|                                    |
         XXXXXXXX|                                    |
        XXXXXXXXX|                                    |
         XXXXXXXX| 1.2                                |
       XXXXXXXXXX|                                    |
 0      XXXXXXXXX|                                    |
         XXXXXXXX| 1.1                                |
          XXXXXXX|                                    |
         XXXXXXXX|                                    |
         XXXXXXXX|                                    |
       XXXXXXXXXX|                                    |
-1       XXXXXXXX|       2.1                          |
           XXXXXX|                                    |
            XXXXX|                                    |
           XXXXXX|                                    |
              XXX|                                    |
               XX|                                    |
               XX|                                    |
               XX|                                    |
=====================================================
```

Fig. 10.12 Item thresholds of two partial credit items

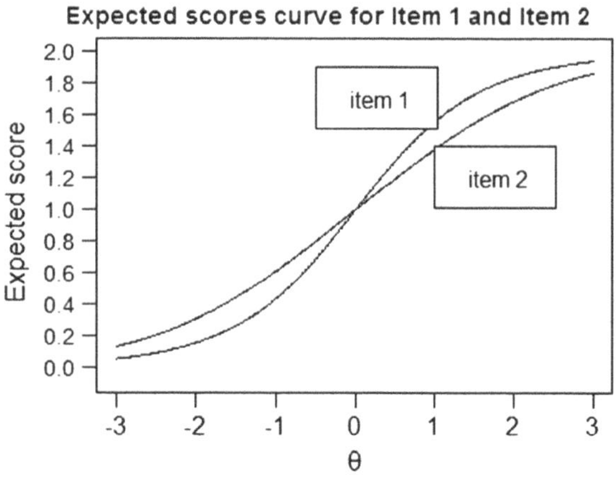

Fig. 10.13 Expected scores curves for two items with different thresholds

difficulty and item discrimination, it should be noted that item difficulty depends on *how many* people obtained the correct answer, and item discrimination depends on *who* obtained the correct answer (i.e. low or high ability students).

Summary

This chapter presents 2PL models for dichotomously scored items and partial credit items. The key difference between 2PL and 1PL (Rasch) models is that item scores are estimated in 2PL, while they are assigned in 1PL. In particular, for 1PL, all item scores are 1 for dichotomous items, and consecutive integer numbers for partial credit items.

There are advantages and disadvantages for choosing either 1PL or 2PL model. An advantage for the 1PL model is that raw scores are sufficient statistics for ability estimates, whether items are dichotomously or polytomously scored. This means that students with the same raw test score will have the same ability estimate. In contrast, with 2PL, because items have different weights in contributing to the ability estimate, students with the same raw test score will likely have different ability estimates, depending on which set of items the student answered correctly. This may pose a difficulty for test administrators to explain to the public.

On the other hand, the 2PL model will always provide better fit to the data. In the case where items do not fit the 1PL model, results from a 1PL analysis are not valid, since the data are not matching the theoretical model. In this case, one may decide to use a model that fit the data better. Generally, 2PL model will results in higher test reliability, since more item level information is taken into account.

In practice, if one aims to achieve better measurement properties in terms of making statements of what students can achieve, and in building learning progressions, one should construct measuring instruments that fit the 1PL (Rasch) model. However, if the item response data do not fit the Rasch model, then it is better to use the 2PL model since that will provide better fit to the model so the results can be consistent with the model being fitted.

Discussion Points

1. Discuss the differences between the Rasch model and the 2PL model. What are the differences in terms of the mathematical formulations of the models? What are the differences in interpreting results of the two models? What are the differences in building measuring instruments?
2. How are the Rasch partial credit model and the 2PL model linked? Why does Chap. 9 say "In fact, when we use the partial credit model and have the task of assigning a weight (maximum score) to an item, we are already thinking in the framework of the two-parameter model?"

Exercises

Q1. In EXCEL, plot ICCs of 2PL items. Vary the difficulty and slope parameters to see the differences. For example

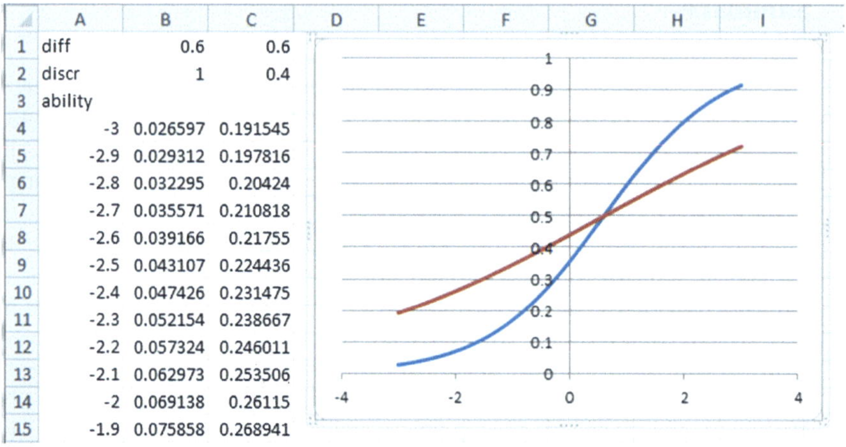

	A	B	C
1	diff	0.6	0.6
2	discr	1	0.4
3	ability		
4	-3	0.026597	0.191545
5	-2.9	0.029312	0.197816
6	-2.8	0.032295	0.20424
7	-2.7	0.035571	0.210818
8	-2.6	0.039166	0.21755
9	-2.5	0.043107	0.224436
10	-2.4	0.047426	0.231475
11	-2.3	0.052154	0.238667
12	-2.2	0.057324	0.246011
13	-2.1	0.062973	0.253506
14	-2	0.069138	0.26115
15	-1.9	0.075858	0.268941

Q2. Indicate whether you agree or disagree with each of the following statements

Item response data fitted with the Rasch model will always have better measurement properties than data fitted with the 2PL model	Agree/Disagree
Residual-based fit statistics are useful for detecting mis-fits of items to the 2PL model	Agree/Disagree
If the **slope parameter** is 0.6 for Item A and 1.2 for Item B, then, under **2PL**, a student who obtained the correct answer for A but incorrect answer for B will have a lower ability estimate than a student who obtained the incorrect answer for A but correct answer for B	Agree/Disagree
If the **item difficulty** parameter is 0.6 for Item A and 1.2 for Item B, then, under **2PL**, a student who obtained the correct answer for A but incorrect answer for B will have a lower ability estimate than a student who obtained the incorrect answer for A but correct answer for B	Agree/Disagree
If the **item difficulty** parameter is 0.6 for Item A and 1.2 for Item B, then, under **the Rasch model**, a student who obtained the correct answer for A but incorrect answer for B will have a lower ability estimate than a student who obtained the incorrect answer for A but correct answer for B	Agree/Disagree
If more students obtained the correct answer to Item A than to Item B, then Item A is likely to be more discriminating than Item B	Agree/Disagree
If a set of items fit the Rasch model so the observed ICCs are approximately parallel, then fitting a 2PL model to the same data will make the observed ICCs criss-cross each other	Agree/Disagree

References

Bock RD (1972) Estimating item parameters and latent ability when responses are scored in two or more nominal categories. Psychometrika 37:29–51

Kiefer T, Robitzsch A, Wu M (2013) TAM (Test Analysis Modules)—an R package [computer software]. http://cran.r-project.org/web/packages/TAM/index.html

Muraki E (1992) A generalized partial credit model: application of an EM algorithm. Appl Psychol Meas 16:159–176

Wu ML, Adams RJ, Wilson MR, Haldane SA (2007) ACER ConQuest version 2: generalised item response modelling software. Australian Council for Educational Research, Camberwell

Chapter 11
Differential Item Function

Introduction

For every IRT model, a mathematical function is used to specify the probability of item responses as a function of the ability (latent trait). The degree to which the observed data fit the mathematical function needs to be examined since valid results can only be drawn if the data fit the model. Chapter 8 discusses the use of fit indices to check for model fit. However, the fit index discussed is only one of many fit indices that can be used to check various kinds of violations to the model. In real-life, data rarely fit a mathematical model precisely, so it is important to carry out such checks.

The term "differential item functioning", or DIF, suggests that an item functions differently in different contexts. As an example, consider a group of students with similar average mathematics abilities for boys and girls. When a mathematics item with a context about baseball is administered to the students, it is found that the boys in this group performed considerably better than girls on this item, even though girls and boys performed similarly on other items. A possible explanation is that boys are more familiar with the question context than girls are, so boys found this item easier than girls did. In this case, we say that the item exhibits DIF for the two gender groups.

In real-life, DIF occurs frequently because every person is an individual and brings his/her own experience and specific knowledge when responding to test items. While the probabilistic nature of item response functions takes into account some variability between individuals, there is often more variability among student responses than what the models accommodate. However, it would not be likely to detect DIF at an individual level, mainly because there is usually insufficient data to do so. DIF is frequently checked for groups such as gender, cultural, geographical and ethnic groups.

The motivation to check for DIF is to ensure that a test is "fair" to all respondents. The inclusion of DIF items in a test will result in lower (or higher) scores for

M. Wu et al., *Educational Measurement for Applied Researchers*,
DOI 10.1007/978-981-10-3302-5_11

some individuals than what they would otherwise obtain should there be no DIF items. Further, DIF also throws some light on the differences among groups of people, such as different strengths and weaknesses between boys and girls. Such information is useful for planning remedial programs.

This chapter explains the meaning of DIF, introduces some detection methods for DIF, and discusses how DIF items should be dealt with.

What Is DIF?

In IRT, the probability of success on an item is a function of a person's ability, θ. Differential item functioning occurs when two groups of people with the same ability, θ, have different probabilities of success on an item. That is, after controlling for ability, the probabilities of success on an item are unequal for the two groups of people. Such items are said to exhibit DIF, and the existence of DIF items would violate the assumptions of the IRT model. The important thing to note here is that the comparison is after "controlling for ability". Suppose two groups of people sat a test. Group A obtained an average score of 30 out of 40. Group B obtained an average score of 20 out of 40. The average scores of the two groups are statistically significantly different, taking into account of the sample sizes of the two groups. In this case, one cannot say that there is DIF just because Group A performed better than Group B, since it is possible that Group A respondents are of higher ability on average than Group B are. Of course it is possible that all items are "biased" against Group B so that Group B performed worse than Group A did. But "the average abilities of the two groups" and "biasedness of the test" are confounded, so that one cannot conclude there is differential item functioning in this test from just examining the overall scores of the two groups. Consequently, DIF is often discussed in relative terms where an item is compared with other items in the test, after overall abilities (or overall test scores) are controlled for, with the assumption that the test as a whole is not biased against any group. As a result, whenever we find items showing DIF favouring one group, we will also find items favouring the other group, since it is assumed that there is no net bias against one group.

Some Examples

OECD PISA 2009 science data have been used as examples to illustrate DIF. The mean science scores for Germany and Taiwan are both 520. If there are no DIF items for these two countries, then one would expect the percentages correct for both countries on each item to be very similar, subject to some random fluctuation due to sampling and measurement errors. Figure 11.1 shows a plot of item-by-item percentages correct for the two countries.

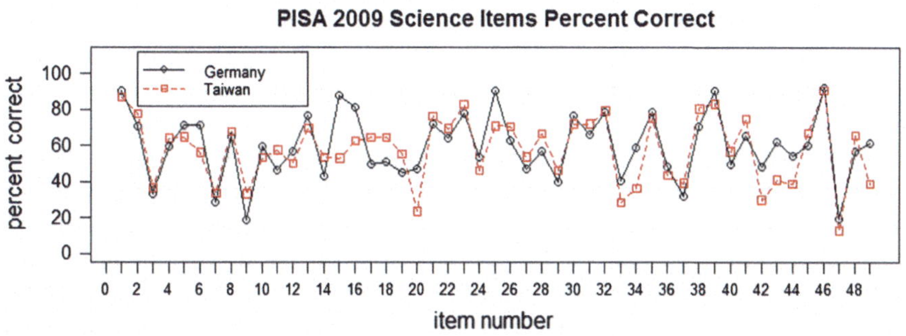

Fig. 11.1 Percentages correct of PISA 2009 Science items for Germany and Taiwan

The first observation about Fig. 11.1 is that, by and large, the percentages correct for Germany and Taiwan are similar for many items. For example, where an item is easy or difficult for Germany (e.g., item 1 or item 3), it is also easy or difficult for Taiwan. That is, the two curves generally move up and down in unison except for a few items. Given that the two countries had the same country mean score, one would expect the two curves to be largely overlapping. However, there are some exceptions. For example, Taiwan students found item 20 more difficult than German students (percentages correct of 46.8 and 23.9% for Germany and Taiwan respectively). On the other hand, Taiwan students found items 17 and 18 easier than German students. Note that although these differences appear to be large visually from the graphs, we still need to carry out formal statistical significance tests to see if these differences are beyond what could be expected due to sampling and measurement errors. Nevertheless, one might conjecture that item 20 shows DIF, given that the difference in percentages correct between Germany and Taiwan is as large as 23%.

A second example shows a comparison of item percentages correct for Japan and Italy on the PISA 2009 Science test. Figure 11.2 shows a plot of item-by-item percentages correct for the two countries.

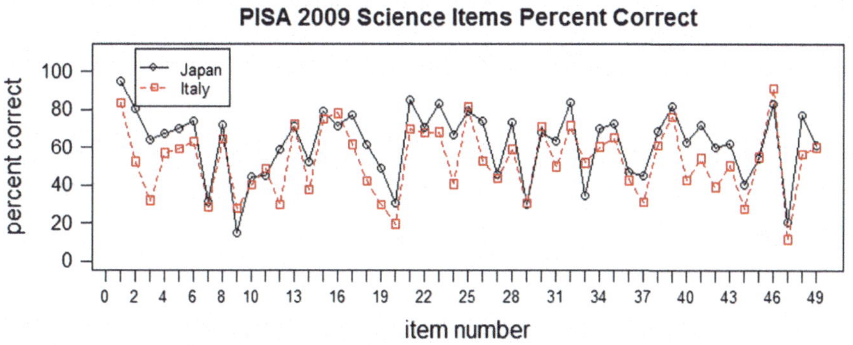

Fig. 11.2 Percentages correct of PISA 2009 Science items for Japan and Italy

Japan's PISA 2009 Science mean score is 539 and Italy's mean score is 489. That is, on average, Japan performed higher than Italy did. As a result, one would expect the percentages correct for Italy to be lower than for Japan for all items. This is largely the case, as Fig. 11.2 shows that the curve for Italy is generally below that of Japan. There are however some exceptions. For example, on a number of items, Italy performed as well as Japan did, or even better than Japan (e.g., items 16, 25 and 46). On the other hand, on some items, Italy did quite poorly as compared with Japan (e.g., items 3, 12 and 24). The fact that Italy performed generally lower than Japan is not an indication of DIF. However, differential differences in percentages correct are signs of possible DIF. That is, if there are no DIF items, one would expect that all percentages correct for Italy to be a little lower than those for Japan. The fact that there are items where Italy performed as well as Japan did, and items where Italy performed a great deal worse than Japan, shows that there are possible DIF items.

Methods for Detecting DIF

There are many statistical methods for detecting DIF. The following provides some examples. These examples are not comprehensive, and many DIF detecting methods are not discussed here. In the literature, there are numerous references on DIF. These include Colvin and Randall (2011), Holland and Wainer (1993), Osterlind and Everson (2009), Zumbo (2007) and Zwick (2012).

Mantel Haenszel

The Mantel-Haenszel test is originally designed for epidemiological studies in which the interest is to study the association between two binary variables while controlling for a confounding variable. More specifically, it studies how stable the strength of relationship is between two binary factors by means of the odds ratios across K different strata that constitute the levels of a confounding variable. The data are presented as a series of 2×2 contingency tables formed by the two binary variables for each value of the confounding variable.

In the 1980s, this test was applied by researchers to study if the status of answering an item correctly or incorrectly is associated with the groups to which the respondents belonged after controlling for their abilities (Holland and Thayer 1988). For example, in studying if there is evidence of differential item functioning (DIF) in a test against the female gender, the female respondents can be regarded as constituting the focal group while the male respondents the reference group. It is usual practice to regard the total number of items correct as the confounding variable, which is frequently collapsed into several, say K, strata that span from the low to the high performance stratum. Respondents from both groups belonging to the same stratum are assumed to have the same ability with respect to the test. The Mantel-Haenszel test is then applied to each item, with gender as one of the binary variables and whether the item was answered correctly as the other binary variable. Since most statistical software has a version of the Mantel-Haenszel test, it has become a popular approach to detect if DIF exists in some items within a test.

If we can assume that there is a common odds ratio between the reference and the focal groups across all strata, then we can use the Mantel-Haenszel test to test if the odds of answering an item correctly for the focal group is the same as that for the reference group. This is the same as testing whether the common odds ratio takes on the value of 1.

Table 11.1 is a typical cross-tabulation that summarizes the performances of two groups with ability at level *k* on item *i*. Let us denote the reference and focal groups as group 1 and 2, respectively. Furthermore, let us denote a correct answer is coded as 1 and a wrong answer as 0. Let A_{ik} represent the number of respondents in group 1 who answered item *i* correctly. Likewise, symbols B_{ik}, C_{ik}, and D_{ik} take on their corresponding meanings. Next, let N_{1ik} and N_{2ik} represent the total number of respondents with ability *k* in the reference and focal group, respectively; and M_{1ik} and M_{0ik} represent the total number of respondents with ability *k* across the two groups who answered the item correctly and incorrectly, respectively. Finally, let T_{ik} denotes the total number of respondents with ability *k* on item *i*.

Under this setting, the Mantel Haenszel common odds ratio can be estimated by using the following formula:

$$\alpha_{MH} = \frac{\sum_k A_{ik}D_{ik}/T_{ik}}{\sum_k B_{ik}C_{ik}/T_{ik}}$$

In the context of a DIF study, the value of the common odds ratio equal to 1 would indicate that there is no DIF for the two groups on item *i*. A value greater than 1 would indicate the item favouring the reference group and a value less than 1 favouring the focal group.

The Mantel Haenszel test is a chi-square test at one degree of freedom which can be computed by the following formula

$$\chi^2_{MH} = \frac{\left[\left|\sum_k A_{ik} - \sum_k E(A_{ik})\right| - 0.5\right]^2}{\sum_k \mathrm{var}(A_{ik})}$$

Table 11.1 Performance table for the two groups with ability at level *k* on item *i*

	1 (correct)	0 (incorrect)	Total
Group 1 (reference group)	A_{ik}	B_{ik}	N_{1ik}
Group 2 (focal group)	C_{ik}	D_{ik}	N_{2ik}
Total	M_{1ik}	M_{0ik}	T_{ik}

where $E(A_{ik})$ and $\mathrm{var}(A_{ik})$ represent the expected value and the variance of A_{ik}, respectively, with

$$E(A_{ik}) = \frac{N_{1ik}M_{1ik}}{T_{ik}}$$

and

$$\mathrm{var}(A_{ik}) = \frac{N_{1ik}N_{2ik}M_{1ik}M_{0ik}}{T_{ik}^2(T_{ik}-1)}$$

As an example, we will use the performance data of students from Germany and Taiwan on science items in Booklet 13 of the PISA 2009 study. For this example, only 18 items were used and the total score of each student from these items was taken as an indicator of the student's ability. The numbers of 15-year old students from Germany and Taiwan working on this booklet amounted to 370 and 461, respectively. Since the sample size was not very large, we collapsed the range of total scores into three strata, spanning from low to high abilities. The analysis was performed using SAS, with Germany's students being regarded as the reference group and Taiwan's students forming the focal group. The common odds ratio for item S256Q01 amounted to 1.8190 and the Mantel-Haenszel chi-square test resulted in $\chi^2(1) = 6.5607$, p = 0.0104. Hence, there is some evidence that this item favoured Germany's students more than Taiwan's students.

There are variations to the above mentioned approach for the purpose of DIF items detection. For example, Holland and Thayer (1988) developed the MH D-DIF index, which is defined as

$$\text{MH D-DIF} = -2.35\ \ln(\alpha_{MH})$$

For our case, the MH D-DIF for item amounted to −1.41, with the negative value indicating that the odds of the focal group in obtaining the correct answer to the item is less than that for the reference group after conditioning on the abilities of the students. There is a set of procedures if one decides to follow the MH D-DIF approach. Interested readers are encouraged to read Holland and Thayer (1988), Dorans and Holland (1993) and other relevant literature for details.

IRT Method 1

The examples given in Figs. 11.1 and 11.2 of this chapter demonstrate that DIF occurs when an item is easier or more difficult *than expected* for a group. A simple way to detect DIF is to carry out an IRT calibration of item difficulties for each group of respondents separately, and then compare the calibrated item difficulties across groups. In these calibrations, we note that the origins of the IRT scales in separate calibrations need to be set to be equal so the calibrated item difficulties are comparable. A simple way is to set the average of the calibrated item difficulties to zero, so the calibrated item difficulties for different groups of respondents are comparable. See Chap. 7 for setting the locations of IRT scales.

Using the Germany-Taiwan data set in the Section "What is DIF?" as an example, we calibrated the item difficulties for Germany and for Taiwan separately. Table 11.2 shows the calibrated item difficulties for the two countries, where the mean of the item difficulties for each country is zero. If there is no differential item functioning, then the item difficulty for an item for Germany should be close to the item difficulty for Taiwan, although these two item difficulties will not be exactly the same as they are subject to measurement error, as reflected in the standard error (s.e.) column in Table 11.2. The sixth column in Table 11.2 shows the differences between the item difficulties of the two countries. A scan down the column shows that some of these differences are rather large (e.g., item 15).

The magnitude of the differences in item difficulties between the two countries can be more easily seen through a scatter plot of the item difficulties, as shown in Fig. 11.3. The solid line in Fig. 11.3 is the identity line (that is, x = y line). It can be seen that while there is a correlation between the item difficulties of the two countries (r = 0.77), some points are far away from the identity line.

Statistical Significance Test

For each item, to see whether the difference between the item difficulties for Germany and Taiwan is larger than expected, a standardised statistic can be computed by dividing the difference by its standard error. Column seven in Table 11.2 shows the standardised statistic, obtained through dividing the difference (column six) by the square root of the sum of squares of the standard errors in columns three and five. The standardised difference in column seven can be regarded as a z-statistic, so that a value outside the range of −2 and 2 can be

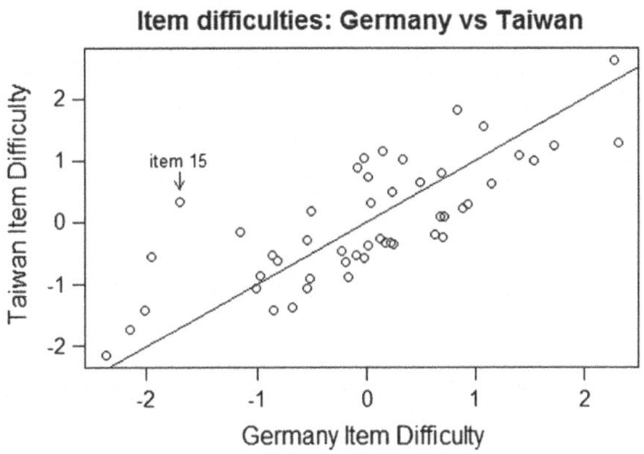

Fig. 11.3 A plot of item difficulties for Germany and Taiwan

Table 11.2 Item difficulties for Germany and Taiwan, calibrated separately

Item	Germany item difficulty	s.e.	Taiwan item difficulty	s.e.	Difference between Germany and Taiwan difficulties	Standardised difference
1	−2.15	0.08	−1.72	0.07	−0.43	−3.78
2	−0.54	0.06	−1.06	0.06	0.52	5.84
3	1.40	0.06	1.07	0.05	0.32	3.97
4	0.13	0.06	−0.27	0.05	0.40	4.91
5	−0.55	0.06	−0.29	0.05	−0.26	−3.06
6	−0.51	0.06	0.17	0.05	−0.69	−8.33
7	1.72	0.06	1.25	0.05	0.47	5.60
8	−0.23	0.06	−0.47	0.06	0.23	2.84
9	2.31	0.07	1.27	0.05	1.04	11.46
10	0.03	0.06	0.31	0.05	−0.27	−3.43
11	0.70	0.06	0.08	0.05	0.62	7.91
12	0.23	0.06	0.48	0.05	−0.25	−3.13
13	−0.87	0.07	−0.54	0.06	−0.32	−3.68
14	0.93	0.06	0.28	0.05	0.65	8.22
15	−1.70	0.08	0.33	0.05	−2.03	−20.89
16	−1.15	0.07	−0.16	0.05	−0.99	−11.09
17	0.70	0.06	−0.25	0.05	0.95	11.76
18	0.63	0.06	−0.21	0.05	0.83	10.49
19	0.88	0.06	0.22	0.05	0.66	8.37
20	0.83	0.06	1.81	0.06	−0.98	−11.77
21	−0.52	0.06	−0.92	0.06	0.40	4.53
22	−0.10	0.06	−0.53	0.06	0.43	5.33
23	−0.85	0.06	−1.42	0.07	0.57	6.21
24	0.49	0.06	0.63	0.05	−0.15	−1.89
25	−1.96	0.09	−0.57	0.06	−1.39	−13.29
26	−0.03	0.06	−0.57	0.06	0.54	6.55
27	0.17	0.06	−0.33	0.05	0.50	6.13
28	0.25	0.06	−0.35	0.05	0.60	7.37
29	1.15	0.06	0.63	0.05	0.52	6.55
30	−0.82	0.06	−0.63	0.06	−0.19	−2.31
31	−0.20	0.06	−0.64	0.06	0.44	5.50
32	−1.01	0.06	−1.06	0.06	0.05	0.54
33	1.07	0.06	1.56	0.06	−0.48	−5.91
34	0.15	0.06	1.15	0.05	−1.00	−12.47
35	−0.97	0.07	−0.87	0.06	−0.10	−1.13
36	0.68	0.06	0.79	0.05	−0.11	−1.45
37	1.54	0.06	1.01	0.05	0.53	6.48
38	−0.68	0.07	−1.38	0.07	0.70	7.47
39	−2.01	0.09	−1.43	0.07	−0.58	−5.19

(continued)

Table 11.2 (continued)

Item	Germany item difficulty	s.e.	Taiwan item difficulty	s.e.	Difference between Germany and Taiwan difficulties	Standardised difference
40	0.68	0.06	0.09	0.05	0.59	7.39
41	−0.17	0.06	−0.90	0.06	0.73	8.47
42	0.01	0.06	0.74	0.05	−0.73	−9.21
43	−0.08	0.06	0.88	0.05	−0.96	−12.09
44	0.33	0.06	1.01	0.05	−0.69	−8.68
45	0.01	0.06	−0.38	0.05	0.39	4.84
46	−2.38	0.10	−2.15	0.08	−0.23	−1.73
47	2.28	0.07	2.62	0.07	−0.34	−3.33
48	0.22	0.06	−0.33	0.05	0.55	6.82
49	−0.02	0.06	1.04	0.05	−1.06	−13.24

regarded as statistically significant. We note that most of the standardised differences in column seven are statistically significant.

Effect Size

One problem with statistical significance test is that when the sample size is large, the data have the power to detect small differences so that statistical tests are likely to show significance. This is because, in real life, the items are likely to work slightly differently across different countries, and the significance test checks if the item difficulties are identical. If the sample size is large, we would nearly always find significant results for all items. Consequently, in addition to testing for statistical significance, one may examine the magnitude of the actual difference and make a judgement of whether the difference is important (similar to the concept of effect size). For example, for item 5, the difference is 0.26 (logit), and the standardised statistic shows significance (3.06). If it is deemed that a magnitude of 0.26 is within acceptable range, then one may ignore the statistical significance test. Of course it is always somewhat arbitrary to choose a cut-off value of logit difference for considering whether DIF exists or not. In our practical experience, often 0.5 logit has been chosen as a cut-off value. But one can always choose a different cut-off value. A practical way to approach this is to begin examining the items with largest DIF, and working back till an adequate set of items are retained.

In this process, take note that first, in real data sets (as opposed to simulated data sets) all items will exhibit DIF when the sample size is large enough. Second, DIF is relative in that an item shows DIF in relation to other items. So after removing DIF items, the comparisons of the remaining items are not quite the same as for the original set of items. That is, there is no notion of "absolute" DIF in that an item will show DIF in comparison to any set of items. Since all items will show DIF when sample size is large enough, one is unlikely to find a set of "DIF-free" items. Therefore, the judgement of the acceptance of a set of items will need to be practical, bearing in mind that we are not likely to eliminate all DIF items.

Having said that, it should be noted that the presence of DIF items can potentially alter respondents' results and lead to manipulations of a test in favour or against a particular group. This is the tension when analysing items for DIF. Such tensions are frequently encountered when tests are constructed.

IRT Method 2

A second method using IRT is to calibrate the data from both groups of respondents together in one calibration instead of separate calibrations. In the IRT model, item-by-group interaction parameters are added to the item difficulty parameters and ability parameters. More specifically, the probability model as shown in Chap. 7,

$$p = P(X = 1) = \frac{\exp(\theta - \delta)}{1 + \exp(\theta - \delta)} \tag{11.1}$$

is modified to include an interaction term, as shown in Eq. (11.2)

$$p = P(X = 1) = \frac{\exp\left(\theta_n - \left(\delta_i + D_{gi}\right)\right)}{1 + \exp\left(\theta_n - \left(\delta_i + D_{gi}\right)\right)} \tag{11.2}$$

where g is the group number, i is the item number and n is the respondent number. Essentially, Eq. (11.2) specifies that the chance of success of a respondent on an item depends not only on the ability of the respondent and the difficulty of an item, but also on an adjustment to the difficulty of the item owing to the membership of the respondent in a group (D_{gi}). For example, if an item favours group 1 respondents, then D_{gi} will be negative, making the overall item difficulty lower than the average difficulty δ_i. On the other hand, if an item is biased against a group of respondent, then D_{gi} will be positive, making the item difficulty higher than δ_i for this group. Note that since the average item difficulty is δ_i, the sum of D_{gi} across the groups will be zero. Further, D_{gi} is an adjustment to item difficulty over and above the overall performance difference between the two groups. Therefore, the sum of D_{gi} across all items for each group is zero, since DIF is not associated with the overall performance difference between the two groups. Consequently, if there are items in a test favouring one group, there will be items in that test favouring the other group. In this way, DIF is measured in the context of the items in a test. If an item appears in a different test, then that item may or may not exhibit DIF.

Equation (11.2) shows an IRT model known as the "facets model", in that the term D_{gi} is a factor influencing the probability of success, in addition to the typical item difficulty and person ability parameters. See Chap. 13 for more information on facets models. In this case, D_{gi} is a term attributable to group membership of the respondents. In other facets models, additional terms in the probability function could be rater harshness, item format or other factors influencing the probability of success, over and above the typically modelled ability and item difficulty. In the context of DIF analysis, D_{gi} is often referred to as "item-by-group" interaction term.

The data set used for IRT method 1 is re-analysed using IRT method 2 where an item-by-country interaction is added to the model. The estimates of this interaction term are given in Table 11.3. In this example, there are two groups (two countries), so the interaction terms are D_{1i} and D_{2i}, where D_{1i} refers to "item-by-Germany" interaction and D_{2i} refers to "item-by-Taiwan" interaction. For model identification, $D_{1i} + D_{2i} = 0$ since the average item difficulty has already been modelled by δ_i in Eq. (11.2). Consequently, $D_{1i} = -D_{2i}$, so that only D_{1i} is estimated, and D_{2i} is set to $-D_{1i}$. Table 11.3 shows estimates D_{1i} and its associated standard errors. In this case, it is the item interaction term for Germany. For example, for item 1, the average item difficulty, δ_1, needs to subtract 0.19 to reflect the item difficulty for Germany, while the item difficulty for Taiwan is $\delta_1 + 0.19$ for this item.

To calculate the magnitude of DIF, the difference between the item difficulty for Germany and item difficulty for Taiwan is computed, that is, $D_{1i} - D_{2i}$. For item 1, this is $(-0.19) - (0.19) = -0.38$. For item 2, the DIF magnitude is $0.26 - (-0.26) = 0.52$. Consequently, the magnitude of DIF is twice the magnitude of the values in the second column (interaction term) in Table 11.3 (i.e., $2D_{1i}$). To carry out a statistical significance test, the magnitude of the DIF is divided by its standard error. Since the magnitude of DIF is $2D_{1i}$, the standard error is also $2 \times s.e(D_{1i})$. Therefore, $\frac{2D_{1i}}{2\times s.e.(D_{1i})} = \frac{D_{1i}}{s.e.(D_{1i})}$. That is, the ratio between D_{1i} and its standard error (columns 2 and 3 in Table 11.3) can be regarded as a z-statistic. Note that in Table 11.3, the standard error for the last item is not available. This is because the interaction term for the last item is not estimated, but set to the negative sum of the other interaction terms so that the average across all items is zero. This is necessary for model identification. The standard error for the last item should have a similar magnitude as for other items (Table 11.3).

To compare IRT method 1 and method 2, a plot of the DIF estimates from the two methods is shown in Fig. 11.4.

It can be seen from Fig. 11.4 that the DIF estimates from IRT methods 1 and 2 are essentially the same. IRT method 2, however, has smaller standard errors for DIF estimates than IRT method 1.

How to Deal with DIF Items?

When DIF items are detected, there are three possible approaches to deal with the items: remove the DIF items, split DIF items as different items for different groups, or do nothing and leave the items in the test. The following discusses each approach.

Table 11.3 Estimates of item-by-country interaction term

Item	Item-by-Country interaction term (Germany)	s.e.
1	−0.19	0.032
2	0.26	0.03
3	0.14	0.028
4	0.19	0.028
5	−0.14	0.029
6	−0.35	0.028
7	0.20	0.029
8	0.11	0.029
9	0.49	0.029
10	−0.15	0.028
11	0.30	0.028
12	−0.14	0.028
13	−0.17	0.029
14	0.30	0.028
15	−1.01	0.029
16	−0.50	0.029
17	0.46	0.028
18	0.40	0.028
19	0.31	0.028
20	−0.52	0.029
21	0.19	0.029
22	0.21	0.028
23	0.30	0.03
24	−0.09	0.028
25	−0.69	0.031
26	0.27	0.029
27	0.24	0.028
28	0.29	0.028
29	0.24	0.028
30	−0.09	0.029
31	0.22	0.028
32	0.03	0.03
33	−0.27	0.028
34	−0.52	0.028
35	−0.05	0.03
36	−0.08	0.028
37	0.24	0.028
38	0.35	0.03
39	−0.28	0.032
40	0.28	0.028

(continued)

Table 11.3 (continued)

Item	Item-by-Country interaction term (Germany)	s.e.
41	0.36	0.029
42	−0.38	0.028
43	−0.50	0.028
44	−0.36	0.028
45	0.19	0.028
46	−0.10	0.034
47	−0.21	0.031
48	0.26	0.028
49	−0.54	NA

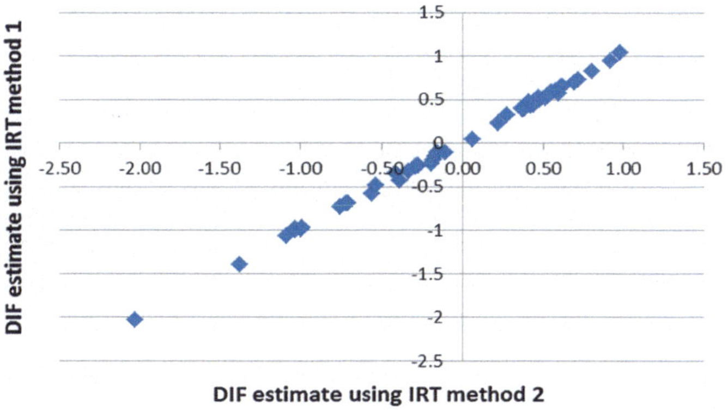

Fig. 11.4 Comparison of DIF estimates using IRT methods 1 and 2

Remove DIF Items from the Test

When there are many test items available for selection into a final test, there is the option of removal of items from a pilot test. Items may be removed because of poor psychometric properties, such as low discrimination, overly easy or difficult items, and items with DIF. Regarding the removing of DIF items, the following points should be considered.

First, the identification of DIF items can be based on statistical significance test or effect size, or both, as discussed earlier. One needs to be aware that statistical significance tests are greatly influenced by sample size, so that more DIF items will be detected when the sample size is large. In real-life, nearly all items will exhibit DIF when the sample is large enough to detect small violations to the model.

Second, DIF is a relative notion. An item exhibits DIF with regard to other items in the test, in very much the same way as item fit, as discussed in Chap. 8. That is, DIF occurs when a group's performance on an item is not as expected based on their performance on other items. So when items are removed from a test owing to DIF, items not showing DIF in the original item set may now show DIF since the item set for checking for DIF is now a different set. Further, if we use 5% level to identify DIF items via significance tests, there will nearly always be items showing statistically significant DIF, in the same way that there are always 5% of people outside the middle 95% region of any distribution. Consequently, one needs to be cautious in using any procedures for trimming off DIF items in stages such as the "purification process" of sequentially removing items and re-scaling to identify a set of DIF items.

Depending on the order of removal of items, the final set of remaining items may vary, because DIF is estimated in reference to the items remaining in the test and different items may be identified as DIF items, should the set of items in the test changes.

For this reason, sometimes there are procedures that identify a set of reference items deemed not to have DIF. Such decisions are made based on substantive reasoning through examinations of the items, and not through statistical tests. Based on the reference set of DIF-free items, we can make judgements about which items are actual DIF items. Consequently, it may not always be the case that items showing statistically significant DIF in favour or against a group are candidate items to be removed. It may be the case that in a test, most items work against a particular group of respondents, or most items are in favour of a particular group, with regard to the reference set of DIF-free items. However, in the absence of a reference set of DIF-free items, we need to make the assumption that the overall test is not biased against or in favour of a group of respondents, but individual items may be. So in choosing items for a final test, one would avoid choosing items predominantly against or in favour of a particular group of respondents.

Split DIF Items as Two New Items

Sometimes DIF analysis is a post hoc procedure after a test has already been administered. Therefore there is no chance of modifying items and re-administering a test, or selecting items to form a new test. In such circumstances, removing DIF items will results in loss of information collected. One way to retain the information collected but at the same time to prevent model violation is to treat DIF items as two different items for two groups of respondents. For example, if item 1 is a DIF item, then this item may be called item 1a for group 1 respondents and item 1b for group 2 respondents. When scaling is carried out, item 1a and item 1b are treated as two different items.

Retain DIF Items in the Data Set

Frequently, when there is no specific set of DIF-free items as a reference, there is an implicit assumption that the overall test is not biased against any particular group. Under this assumption, retaining DIF items will not significantly change the estimated abilities of the respondents, because there will be DIF items favouring as well as against each group of respondents so that it "evens out". For this reason, when there are no DIF-free items as a reference set, DIF tests will not be able to detect "real bias" of a test. DIF only detects item-level differences with the assumption that there is no overall bias.

While the ability estimates will not change a great deal when DIF items are retained, the descriptions of skills progression will be inaccurate. That is, the estimated difficulty of a DIF item will not reflect the real difficulty for either group of respondents. In IRT method 2, we need to add the adjustment term, D_{gi}, to obtain the item difficulty for each group.

One rationale for retaining DIF items in a test is that it is a fact of life that DIF exists, and removing these items will not reflect what is happening in the real-world. In fact, some argue that DIF items provide useful information for teaching and learning, since teaching programs can be tailored for different groups of respondents, after finding out relative strengths and weaknesses of student groups. For example, in a study it has been found that girls generally do not perform as well as boys in spatial mathematics while they perform better than boys in the numbers strand of mathematics. In reading, boys perform as well as girls in factual texts but less well in making inferences from texts. If tests do not contain DIF items, then these differences will not be found.

Cautions on the Presence of DIF Items

Having said that there are merits in not removing DIF items in tests, there are some cautions regarding having DIF items in tests. Clearly, when items have DIF, it becomes possible to manipulate test compositions to favour or to be against a particular group. In the example above, it is possible to manipulate the compositions of different text types in a reading test so that either boys or girls will perform relatively better (or worse) than they would, had there been a balanced composition of text types. That is, the existence of DIF items opens up the possibility of manipulation of test results. Consequently, all tests should undergo not only DIF analysis through statistical procedures, but also substantive content analysis by experts to determine whether a test may be biased against a particular group.

Further, when test results hinge on a small set of test items, such as equating between tests using link items, or state-wide tests using only around 40 items, it is crucial that item bias due to DIF is carefully examined. If many of these items have DIF, they could have considerable influence on the results. Because the number of items is small, DIF effects in favour and against a group is not likely to "even out". Wu (2010) showed that DIF exists in PISA and DIF items used for linking between PISA cycles can significantly change country mean scores in PISA.

A Practical Approach to Deal with DIF Items

For a practical approach to deal with DIF items, the first step is to use statistical analysis to identify DIF items. DIF items should not be automatically deleted without an examination of item content to look for theoretical explanations for the presence of DIF. Is the item testing the construct intended? For example, if migrant students find a mathematics word-problem difficult, it could be because of a lack of language skills than a lack of mathematics skills on the part of the students. Is this intended? Is the ability of reading in the construct being tested? On the other hand, if girls do not perform as well as boys do for a spatial mathematics problem, but spatial ability is part of the mathematics construct being assessed, then one may want to retain the item.

Clearly, items with very large DIF are candidates for deletion, for example, item 15 in Table 11.2 where the difference in item difficulties for the two countries is more than 2 logits. Should this item be selected for equating purposes for future cycles of PISA, it could have a significant effect on results for the two countries. A closer examination of this item will be helpful.

More generally, it is advisable to begin examining items with the largest DIF magnitudes, and eliminate items or split items as necessary, and progressively work back for items with decreasing magnitudes of DIF. But do not expect to have a final set of DIF-free items. The "stopping rule" for accepting items with some DIF is probably best decided by practicality, that is, in the end, you need to have sufficient number of items to run a test, but bearing in mind that DIF items can distort results. Overall, statistical procedures and substantive reasoning should be used together to deal with DIF items.

Summary

This chapter explains the concepts of differential item functioning (DIF), presents some methods for detecting DIF, and provides suggestions on how to deal with DIF items.

DIF occurs when respondents with the same ability have different probabilities of success on an item. DIF is caused by different strengths and weaknesses of respondents owing to a number of possible factors, including different curriculum, different personal disposition, experience, culture, language and many other reasons. It is a fact of life that DIF exists in many tests for different groups of respondents. There are many statistical procedures to detect DIF. Three methods are introduced and compared in this chapter. The detection of DIF does not mean that DIF items are necessarily problematic, but cautions must be taken when tests are short, or when link items are selected for equating purposes to ensure that any presence of DIF does not threaten the assessment results. It is noted that statistical procedures for the identification of DIF items should be complemented with theory-driven explanations of the occurrence of DIF. Items should not be removed based on statistical decisions alone. An understanding of the reasons for DIF is also essential for achieving fair tests.

We note that the DIF detection methods introduced in this chapter only focuses on item difficulty parameters. This is the case for the Rasch model where item difficulty parameters are estimated. For 2PL IRT models where discrimination parameters are also estimated, DIF detection may involve both the item difficulty and discrimination parameters. That is, an item may have different discrimination power for two groups of respondents, in addition to different item difficulty parameters. This is beyond the scope of this book.

Hands on Practise

(1) The following table shows the percentages correct of PISA 2009 mathematics items for Shanghai and Australia (Table 11.4).

The following shows a plot of these percentages correct.

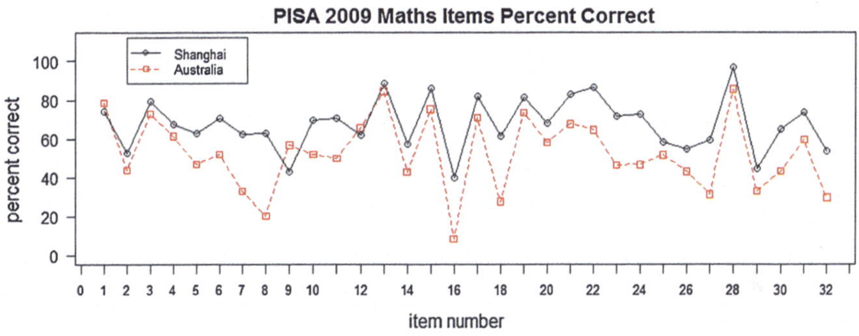

Given the information provided in the table and the graph, discuss the relative performances of Shanghai and Australian students, and give your impressions of the presence/absence of DIF.

Discussion Points

(1) Why is it that the existence of DIF items would violate the assumptions of the IRT model applied to the test?
(2) Do you agree that the ability estimates will not change a great deal when DIF items are retained? Why or why not?

Table 11.4 Percentages correct for Shanghai and Australia PISA Mathematics items

Item	Percentage correct		
	Shanghai	Australia	Difference
M033Q01	74.1	78.6	−4.5
M034Q01T	52.5	43.8	8.7
M155Q01	79.4	73.2	6.3
M155Q04T	67.7	61.5	6.2
M192Q01T	63.0	47.1	15.9
M273Q01T	70.3	52.3	18.0
M406Q01	62.3	33.4	29.0
M406Q02	62.9	20.5	42.3
M408Q01T	42.9	57.1	−14.2
M411Q01	69.4	52.0	17.3
M411Q02	70.4	50.1	20.3
M420Q01T	61.8	65.7	−4.0
M423Q01	88.5	84.6	3.9
M442Q02	57.1	43.0	14.1
M446Q01	86.1	75.3	10.7
M446Q02	40.0	8.7	31.3
M447Q01	81.9	71.1	10.8
M464Q01T	61.4	27.8	33.6
M474Q01	81.3	73.7	7.6
M496Q01T	68.2	58.2	10.0
M496Q02	82.6	67.8	14.8
M559Q01	86.2	64.7	21.5
M564Q01	71.5	46.4	25.1
M564Q02	72.5	46.8	25.7
M571Q01	58.2	51.9	6.3
M603Q01T	54.6	43.3	11.3
M603Q02T	59.3	31.5	27.8
M800Q01	96.8	85.6	11.2
M803Q01T	44.3	33.1	11.2
M828Q01	64.8	43.5	21.2
M828Q02	73.5	60.0	13.6
M828Q03	53.6	29.9	23.7
Average	67.5	52.6	14.9

Exercises

Q1. Indicate whether you agree or disagree with each of the following statements

A statistical DIF analysis did not detect any DIF item in a test. We can be assured then the test is a fair test for the groups of respondents for which the DIF analysis was performed	Agree/disagree
To determine if a test is a fair test, count the number of DIF items in a test using a statistical DIF detection procedure. If there are considerably more items favouring one group than the other group, then the test is biased	Agree/disagree
A test is biased if the average performance of one group of respondents is considerably higher than for other groups	Agree/disagree
When sample size increases, the magnitudes of DIF (difference in item difficulties for two groups of respondents) will tend to increase	Agree/disagree
When sample size increases, more items will tend to show statistically significant DIF estimates	Agree/disagree

References

Colvin KF, Randall J (2011) *A review of recent findings on DIF analysis techniques* (Center for Educational Assessment Research Report No. 795). University of Massachusetts, Amherst, Center for Educational Assessment, Amherst, MA

Dorans NJ, Holland PW (1993) DIF detection and description: Mantel-Haenszel and standardization. In: Holland PW, Wainer H (eds) Differential item functioning. Lawrence Erlbaum Associates, Hillsdale, NJ, pp 35–66

Holland PW, Thayer DT (1988) Differential item performance and the Mantel-Haenszel procedure. In: Wainer H, Braun HI (eds) Test validity. Lawrence Erlbaum Associates, Hillsdale, NJ, pp 129–145

Holland PW, Wainer H (1993) Differential item functioning. Lawrence Erlbaum, Hillsdale, NJ

Mantel N, Haenszel W (1959) Statistical aspects of the analysis of data from retrospective studies of disease. J Natl Cancer Inst 22:719–748

Osterlind SJ, Everson HT (2009) Differential item functioning. Sage Publishing, Thousand Oaks, CA

Wu ML (2010) Measurement, sampling and equating errors in large-scale assessments. Educ Measur Issues Pract 29(4):15–27

Zumbo BD (2007) Three generations of differential item functioning (DIF) analyses: considering where it has been, where it is now, and where it is going. Lang Assess Q 4:223–233

Zwick R (2012) *A review of ETS differential item functioning assessment procedures: flagging rules, minimum sample size requirements, and criterion refinement* (ETS Research Report No. RR-86-31). ETS

Chapter 12
Equating

Introduction

When different tests are administered, the results from the tests are not directly comparable. A process called Equating is needed for comparing results from different tests. We first explain the reasons why test results are not directly comparable, followed by the presentation of some equating procedures as examples.

When a person performs well on a test, we do not know whether it is because the person is very able, or because the test is easy. All we can say is that the person found a particular test easy. That is, the test is easy relative to the person's ability. To make statements about a person's ability level in more *absolute* terms (than relative terms), we need some external references and contexts, such as the tests are judged (e.g., by content experts) to be at certain standards (e.g., suitable for 10 years old students), or the person's standing in a reference group (e.g., of the Grade 5 students) so that high or low ability can be interpreted in a context rather than that a respondent just did well (or poorly) on a test.

Mathematically, we can see this confounding between person ability and item difficulty, as in Eq. (7.1)

$$p = P(X = 1) = \frac{\exp(\theta - \delta)}{1 + \exp(\theta - \delta)} \quad (12.1)$$

where the probability of success is a function of the difference between ability and item difficulty. The probability is high when the difference is large. So when a person is successful on an item, we can postulate that the person's ability is likely higher than the item difficulty, but we do not know the actually values of ability and item difficulty in any absolute terms, as both can be somewhat higher or lower.

When ability is conceived as a quantity on a line, where the line goes from negative infinity to positive infinity ($-\infty$ to $+\infty$), the line has no labels on it to indicate where zero is, or how wide a unit on the line might mean. That is, the

M. Wu et al., *Educational Measurement for Applied Researchers*,
DOI 10.1007/978-981-10-3302-5_12

ability line has no lower bound or upper bound, nor any particular unit, other than for comparing relative high and low between people, between items, and between people and items. Therefore, when estimating respondents' abilities and item difficulties, we can use any part of the line to model item difficulties and person abilities. To avoid the problem of obtaining multiple possible estimates of abilities and item difficulties, an arbitrary location on the line is often chosen as the zero point, and the scale unit is also arbitrarily chosen when estimating abilities and item difficulties. Technically, this is referred to as model identification by setting constraints to the solutions of the estimation equations.

As a convention, the average item difficulty across all items or the average ability of all respondents can be set as the zero point of the ability scale. This sets the point of reference for the locations on the line representing ability and item difficulty measures. For the unit of scale of the ability line, it is typically set by assigning "1" to the scale parameter of the Rasch model. That is, in the more general form of the item response function (i.e., the 2PL model discussed in Chap. 10)

$$p = P(X = 1) = \frac{\exp(a(\theta - \delta))}{1 + \exp(a(\theta - \delta))} \tag{12.2}$$

the scale parameter, a, is set to 1, so Eq. (12.2) become Eq. (12.1) for the Rasch model. Setting a to 1 fixes the unit of the ability scale. There is no particular reason why a needs to be set to 1. It can be set to any number. In the Hands-on practice of Chap. 7, we have shown that sometimes a is set to 1.7. For 2PL models, the unit of the scale can be set by fixing the variance of the abilities of respondents to 1 (or to any fixed number), or by setting the sum of the scale parameters a_i across all items to a fixed number, such as the maximum possible score on the test. In this way, the location and scale of the ability line can be (arbitrarily) fixed and a unique set of parameter estimates can be obtained.

Given that the location and scale of the ability measures are arbitrarily fixed, the comparability of the results from two test administrations depends on how the location and scale are determined. For example, if we set the mean of the item difficulties of each test to zero to define the location of the scale, then the zero locations from these two tests are not comparable if the two tests do not have identical items. So to align the locations of measures from two tests, we need to at least have some common items across the two tests to give us a basis for comparison. Consequently, it is really important to realise that if any analyses are to be made to measure trends in student performance or growths, one must design the test instruments to take into account of the need of common items or other mechanisms for equating tests. If equating has not been well planned before test administration, it may well be impossible to link between tests and make comparisons of test results.

Overview of Equating Methods

In this section, we explain the rationales behind equating methods. Our focus in this chapter is on equating the locations of ability scales, and not so much on equating the scale factor of ability scales. Under the Rasch model, there is an assumption that all items in a test have the same discrimination power [i.e., the same a parameter in Eq. (12.2)]. If this assumption is met, there is then no need to equate the unit of the ability scale, since the unit should be the same if the tests under comparison all measure the same latent trait. In practice, this assumption is of course not likely to be met. We discuss the violations of this assumption later in this chapter. We also stress that the equating methods presented in this book are by no means comprehensive. For a more complete discussion on equating methods, see Kolen and Brennan (2004).

Common Items Equating

If two tests have some common items, then the common items can be used to align the two tests. Common items in tests are also known as link items. Formally, such a design is known as common-item non-equivalent groups design (Kolen and Brennan 2004) where the two tests containing common items are administered to different groups of respondents. There are a number of ways to perform equating using common items. We discuss the shift method, anchor method and joint calibration method. However, before the common items are used for linking tests, the items must be checked that they perform *in the same way* in both tests (termed item invariance in this chapter). It is possible that an item may change across different test administrations. For example, if curriculum has changed, the change could affect the item difficulty of an item over time. If DIF exists (see Chap. 11), an item may have different difficulties for two groups of respondents. So even two items are identical, they may not have the same difficulty in different test administrations. Consequently, before using common items for equating, the items should first be checked for item invariance, as described below.

Checking for Item Invariance

We use an example data set to illustrate the procedures for checking for item invariance in terms of item difficulty measures. In this data set, two different tests containing common items were administered three years apart. We use the term *common items* and *link items* interchangeably in this chapter. The tests were calibrated independently. We will use the term "free calibration" to refer to the estimation of item parameters based on a test alone and not linked to any other test

Table 12.1 Item difficulties of common items

Item number	2010 item difficulty	2013 item difficulty
1	−1.392	−1.835
2	−0.189	−0.586
3	−1.087	−1.216
4	−0.812	−1.285
5	0.599	0.666
6	−0.664	−0.952
7	0.837	0.561
8	−0.406	0.514
9	0.641	0.634
10	−0.918	−0.937
11	−0.93	−0.879
12	1.467	1.435
13	−0.393	−0.472
14	−0.766	−0.904
15	−1.464	−1.88
16	−0.566	−0.901
17	−0.798	−1.219
18	0.13	0.418
19	−1.109	−1.085
20	−0.133	−0.164
21	0.888	0.945
22	−2.626	−2.874
23	1.252	0.136
24	0.982	0.45
25	−0.459	−0.45
26	−0.312	−0.51
27	−1.411	−1.891
28	0.348	0.3
29	−0.356	−0.417
30	−1.693	−1.709
31	−0.746	−0.909
32	−0.665	−0.748
Average	−0.398	−0.555

results. Table 12.1 shows the item parameters of the common items when the tests were calibrated "free".

Two observations can be made from Table 12.1. First, the average item difficulties of the common items from the two tests are not the same (−0.398 and −0.555 respectively). This may happen, for example, if the calibration of each test makes the average item difficulties zero as a constraint, then the average of the common items may not be zero, as it depends on what other items are in the tests. Second, the relative difficulties of the common items across the two administrations

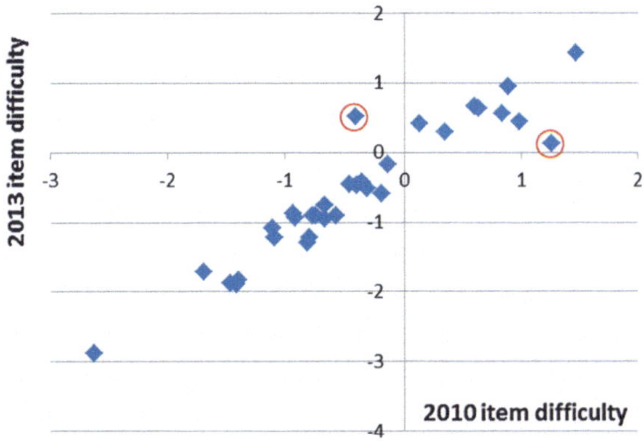

Fig. 12.1 2010 item difficulties versus 2013 item difficulties for the common items

are correlated, as can be seen from a scatter plot of the two sets of item difficulties. Figure 12.1 shows a plot of the 2010 item difficulties against the 2013 item difficulties for the common items.

Generally, difficult items in 2010 are also difficult in 2013, and easy items are easy in both test administrations. However, two items appear further away from the general trend line in Fig. 12.1, namely, item 8 and item 23 (circled). That is, a visual inspection of the performance of the common items indicates that item 8 and item 23 may not be invariant in that the item difficulties may have changed from 2010 to 2013.

Visual checks of item invariance should always be carried out, as it provides a check and assurance of whether the common items are suitable for equating purposes. The visual invariance check can be refined to formally test for statistical significance for the invariance of item parameters.

As the average item difficulties of the common items are not equal across the two test administrations, the magnitudes of the item difficulties are not directly comparable other than through plotting a scatter graph. To formally check the invariance of item parameters, the first step is to align the item parameters so that the 2010 and 2013 sets of common items have the same average value. This can be done by adding the difference between the averages $-0.398 - (-0.555) = 0.157$ to the 2013 estimates. That is, the average for the 2013 common items is lower than the average for the 2010 common items by 0.157. So if all 2013 common items are added by 0.157, the average of the 2013 common items will be the same as the average for the 2010 common items. Table 12.2 shows the adjusted 2013 item difficulties for common items (column 3 in the table) to the 2010 scale where every 2013 item difficulty in Table 12.1 is added 0.157 which is the difference between the average item difficulties in Table 12.1.

In Table 12.2, the difference between the 2010 and 2013 item difficulties (column 4 in table) are computed, after placing the 2013 item difficulties on the 2010 scale. Further, a standardised difference can be computed by dividing the difference in difficulties by its standard error. The standard error for the difference is computed as the square root of the sum of squares of the standard errors for the 2010 and 2013

Table 12.2 Item difficulties of common items with 2013 difficulties on 2010 scale

Item number	2010 item difficulty	2013 item difficulty adjusted	Difference	S.E. 2010	S.E. 2013	Standardised difference
1	−1.392	−1.678	0.286	0.034	0.039	5.53
2	−0.189	−0.429	0.240	0.032	0.036	4.99
3	−1.087	−1.059	−0.028	0.033	0.037	−0.56
4	−0.812	−1.128	0.316	0.032	0.037	6.47
5	0.599	0.823	−0.224	0.032	0.037	−4.57
6	−0.664	−0.795	0.131	0.032	0.036	2.73
7	0.837	0.718	0.119	0.032	0.037	2.44
8	−0.406	0.671	−1.077	0.027	0.029	−27.17
9	0.641	0.791	−0.150	0.032	0.037	−3.06
10	−0.918	−0.780	−0.138	0.032	0.037	−2.81
11	−0.93	−0.722	−0.208	0.032	0.036	−4.31
12	1.467	1.592	−0.125	0.034	0.041	−2.34
13	−0.393	−0.315	−0.078	0.031	0.036	−1.63
14	−0.766	−0.747	−0.019	0.032	0.037	−0.38
15	−1.464	−1.723	0.259	0.034	0.04	4.94
16	−0.566	−0.744	0.178	0.032	0.036	3.70
17	−0.798	−1.062	0.264	0.032	0.037	5.40
18	0.13	0.575	−0.445	0.025	0.029	−11.61
19	−1.109	−0.928	−0.181	0.032	0.037	−3.69
20	−0.133	−0.007	−0.126	0.031	0.036	−2.64
21	0.888	1.102	−0.214	0.032	0.038	−4.30
22	−2.626	−2.717	0.091	0.038	0.046	1.53
23	1.252	0.293	0.959	0.034	0.036	19.37
24	0.982	0.607	0.375	0.033	0.036	7.69
25	−0.459	−0.293	−0.166	0.031	0.035	−3.54
26	−0.312	−0.353	0.041	0.031	0.035	0.88
27	−1.411	−1.734	0.323	0.034	0.04	6.16
28	0.348	0.457	−0.109	0.032	0.036	−2.26
29	−0.356	−0.260	−0.096	0.032	0.035	−2.02
30	−1.693	−1.552	−0.141	0.034	0.039	−2.72
31	−0.746	−0.752	0.006	0.032	0.036	0.13
32	−0.665	−0.591	−0.074	0.032	0.036	−1.53
Average	−0.398	−0.398				

item parameters. For example, for item 1, the standardised difference is computed as $\frac{0.286}{\sqrt{0.034^2 + 0.039^2}} = 5.53$.

The standardised differences can be regarded as a z-score, so that values outside the range of −2 to 2 are statistically significant. As for the discussions on DIF about statistical significance and effect size, when sample size is large, most standardised differences shown in Table 12.2 will be statistically significant. Thus one may look for *outliers* such as items 8, 18 and 23 where the standardised differences are much larger than others, rather than where the standardised difference is outside the −2 and 2 range. Nevertheless, deciding on the items to remove from the common item pool is a somewhat subjective process. The results are also likely to differ if items are removed progressively and scaling is re-run after each item removal, similar to the discussions about DIF item removals in Chap. 11. In practice, we have often tried a number of different criteria for the removal of items from the common item set and compared the results. It may also be helpful to have different data analysts to independently select items as link items for the purposes of equating, and then compare the results. The bottom line is that there is no *one correct way* of selecting link items.

Number of Common Items Required for Equating

When the number of common items is few, the removal of a few items for linking (or the selection of link items) will have considerable effect on the equating results. We recommend a minimum of 30 link items for equating purposes, and more if possible. If the purpose of equating is to put different grade level students' results on the same scale, one should be aware that a yearly increase (i.e., between two adjacent grades of students) in proficiency is of the order of 0.5 logit (about half a standard deviation of the student ability distribution for a year level) based on our experience in dealing with educational type of data (e.g., see Wu 2010). If the purpose of equating is to monitor trend from a calendar year to another calendar year for students in the same grade, one needs to be aware that the average cohort change across time is typically very small (perhaps less than 0.05 logit, or less than one month's of growth), so that we need very accurate measures of mean scores. Errors due to equating are often too large for the purposes of measuring trends. Many link items are needed in that case.

Factors Influencing Change in Item Difficulty

As discussed in previous sections, item difficulty for an item can change owing to curriculum change, exposure of the item and many other reasons. One factor that impacts on item difficulty is the item position in a test. An item placed at the beginning of a test will likely be easier than the same item appearing at the end of a test. Even for a 60-minute test, the so-called fatigue effect is often present. So if a link item is placed at the beginning of one test, while it is placed at the end of another test, the item will likely have different item difficulties. Thus if this item is used as a link item, it will threaten the equating of the two tests. The item position effect has been discussed extensively in Chap. 3. When equating is undertaken, the test design should take into account of item position effect and how test design can mitigate some of the problems. In particular, rotated test booklet design can allow an item to appear in different positions in different test booklets. If rotated test booklets design is not used, then link items should be placed in matching positions in different tests.

Shift Method

In the previous section on procedures for item invariance checks, it is mentioned that the 2013 item parameters are placed on the 2010 scale by computing the difference in the means of item parameters for 2010 and 2013 common items, and adding the difference to all 2013 item parameters. Therefore, one simple method of equating is the method of *shift*, where two tests are calibrated separately, and the item and ability parameters for one test is placed on the scale of another test by a constant shift, the magnitude of which is computed as the amount needed to make the means of the common item parameters between the two tests the same.

In the above example, if the three problematic items are removed from the common item set (i.e., items 8, 18, 23) after the item invariance checks, the mean of the 2010 common items is −0.473, while the mean of the 2013 common items is −0.649. Therefore, to equate the 2013 results onto the 2010 scale, all item parameters (not just the common item set) and all ability parameters need to be added the value of 0.176 (= −0.473 − (−0.649)). That is, all 2013 results are *shifted* by a constant (0.176 in this case).

Mathematically, we may represent the equating transformation as

$$\begin{aligned} T_2' &= T_2 + c \\ &= T_2 + (\mu_1 - \mu_2) \\ &= (T_2 - \mu_2) + \mu_1 \end{aligned} \tag{12.3}$$

where T_2 represents the calibrated (item and ability) parameter values of test 2. The symbol c represents the constant to shift the T_2 values, where c is equal to the difference between the mean values of calibrated common item sets (mean of test 1 common items minus mean of test 2 common items, $\mu_1 - \mu_2$). T_2' represents the transformed parameter values for test 2. Mathematically, we can think of the transformation as a two-step process. First, we subtract the mean of test 2 common set items from T_2 (so that the common item set for test 2 will have a mean of zero), and, in the second step, we add the mean of test 1 common item set to the result from step 1 (so that the mean of the common item set for test 2 will equal to the mean of common item set for test 1).

Shift and Scale Method

The method described above shifts the locations of items and respondents. The method does not take into account of whether the standard deviations of the parameters of the common item set are equal for the two tests. That is, whether the set of items provides the same discriminating power for both tests. Under the assumption that the items of the two tests fit the Rasch model and the two tests are testing the same construct, the standard deviations of the common item sets for both tests should be equal, so one may argue that there is no need to use any multiplying factor to transform the parameters, and that adding a constant to the parameters is sufficient for equating two tests. Since this is an assumption, this assumption ought to be checked in any case. For our example, the standard deviation of the common set of items for 2010 is 0.91, and 0.97 for 2013. This suggests that the 2013 set of items may spread out the respondents more than the 2010 set of items do. That is, in addition to a shift of the item parameters, the ability scale may need to be multiplied by a scale factor to spread out the respondents more, or less.

A variation to the shift method is to transform both the scale and the location of the item parameters. In the case where the standard deviations of the common items for two tests are different and are taken into account in matching test 2 to test 1, the equating transformation is

$$T_2' = \frac{(T_2 - \mu_2)}{\sigma_2}\sigma_1 + \mu_1 \tag{12.4}$$

We can think of Eq. (12.4) as a four-step process. First, adjust test 2 parameters by subtracting the mean of the common item set of test 2. Second, divide the result by the standard deviation of test 2 common items. The combined effect of the first two steps is equivalent to computing a z-score (standardised score) that has a mean of 0 and a standard deviation of 1. The third step multiplies the standard score by the standard deviation of test 1 common items, and the fourth step adds the mean of test 1 common items. The third and fourth steps convert the previously computed z-score to a statistic with mean of μ_1 and standard deviation of σ_1.

Note that we have not provided guidelines as to when Eq. (12.3) should be used and when Eq. (12.4) should be used. In the case of our example where the standard deviations differ somewhat ($\sigma_1 = 0.91, \sigma_2 = 0.97$), it is not immediately clear which equating transformation should be used. The standard deviations of the common item sets can be quite sensitive to outlying observations, (or extreme values) particularly when the number of common items is small (say, fewer than 30). For example, in our example, if the lowest item parameter in test 2 is −2.2 instead of the observed −2.9, the standard deviation will change from 0.97 to 0.93. As mentioned in previous sections, item position effect alone could change item parameter values considerably, and in turn affect the mean and standard deviation of the common item sets.

In practice, different equating methods should be carried out and decisions can be made after comparing the results from different approaches. Unfortunately, in real-life, data are often *messy*, and it is not always clear which analysis method is the best one to apply. This is particularly relevant in equating tests. In one of our projects, seven different equating methods produced seven different sets of results. The instability of the results can be traced back to poor test design in terms of the placement of items and the number of common items. Consequently, careful planning at the test preparation stage is essential for achieving valid and reliable results in making inferences across multiple tests.

Shift and Scale Method by Matching Ability Distributions

Another variation to the shift and scale method is to match the ability distributions rather than to match the item parameter distributions. For the example given above, two tests with common items are administered in 2010 and 2013. IRT scaling is carried out for each test separately so that two sets of item parameters are produced. A re-scaling of the 2010 data using 2013 item parameters for the common items is carried out, where the item parameters are fixed at the 2013 calibrated values. The result of this new 2010 calibration produces ability estimates different from the original 2010 calibration, since different item parameters are used. Let μ and σ denote the mean and standard deviation of the ability distribution from the original 2010 calibration, and let μ' and σ' denote the mean and standard deviation of the ability distribution from the second calibration where 2013 item parameters are used to calibrate 2010 data. Since, 2010 data are used in both calibrations, the mean and standard deviation from both calibrations should be matched. The following shows a transformation that can make the mean and standard deviation of the ability distribution from the second calibration match that of the original calibration of the 2010 data.

$$T' = \frac{(T - \mu')}{\sigma'}\sigma + \mu \tag{12.5}$$

where T is any calibrated parameter using the 2013 item parameters. Equation (12.5) is very similar to Eq. (12.4), where a z-score is first formed ($\frac{(T-\mu')}{\sigma'}$), and then the z-score is transformed to have a standard deviation of σ and mean μ. Once this transformation is derived, we can apply the transformation to the 2013 calibration. That is, Eq. (12.5) is applied to all 2013 calibrated item and person parameters to place them on the 2010 scale.

Anchoring Method

Another way to place 2013 results on the 2010 scale is to use 2010 item parameters for the common items when 2013 data are scaled. That is, when scaling the 2013 data, the item parameters for the common items in 2013 are not estimated but fixed (or known as anchored) on the values from the 2010 scaling. All other items that are unique to the 2013 test are scaled relative to the common item set, so these new items will be placed on the 2010 scale. Note that in this case, no other constraint should be placed. That is, we should not set the mean item difficulty or mean ability to zero, as these constraints may not be compatible with constraining the common items to fixed item parameters.

The anchoring method differs from the shift method in that every common item in the 2013 test is fixed to the 2010 parameter value, while in the shift method, only the mean of the set of 2013 common items are made equal to the mean of the 2010 common set so that individual items in the common item set may not have equal parameter values. Consequently, the anchoring method applies more stringent constraints. The shift method allows for some leeway for the item parameters to change between two test administrations, as long as the set of common items have the same mean value across the two tests. This may be a more realistic scenario where items may have small changes between test administrations, as item positions in tests may not be identical. There could also be many other reasons why item parameters may not be identical between two test administrations. Typically, the two tests being equated have different cohorts of students who might bring with them different background knowledge. From these points of view, the shift method would seem preferable to the anchoring method, as it allows for some variation in item parameters.

However, there are situations where the anchoring method may be preferred to the shift method. For example, if the 2013 test has a smaller sample of respondents and the calibration of 2013 data is deemed less reliable, while the 2010 item parameters have been well established, it may be preferable to use the 2010 item parameters as anchors when calibrating 2013 data. In practice, the shift and anchor methods are not expected to make a great deal of difference to the results. What is important is the number of common items and the invariance characteristic of the common items. Strictly speaking, if the items are invariant across two test administrations, then the shift and anchor methods should produce very similar results.

The Joint Calibration Method (Concurrent Calibration)

Another easy way to equate between two tests with common items is to perform a joint calibration of both sets of test data. This is also known as concurrent calibration. That is, the data from both tests are combined, with the common items aligned in the combined data set, as illustrated by Fig. 12.2.

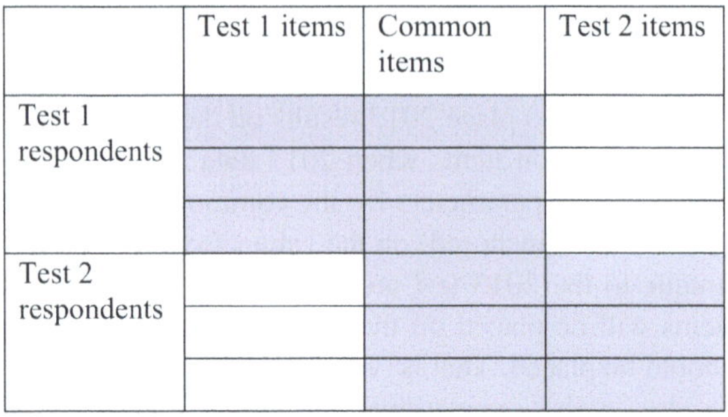

Fig. 12.2 Arrangement of data from two tests with common items for joint calibration

To combine the data sets from two tests with common items, append one data set to another data set as shown in Fig. 12.2, but make sure that the data for the common items are matched so that both data sets contribute to the response data of the common items. The shaded areas in Fig. 12.2 show the item responses. For the items unique to each test (i.e., not common items), there will be missing responses in the other tests (the unshaded spaces in Fig. 12.2). When data are combined as shown in Fig. 12.2, the calibrated item parameters for the common items will be based on item responses from both data sets, while the unique items in each data set are calibrated in relation to the common item sets, so that the unique items from both tests are also on the same scale, as these items are all linked to the common items.

While the joint calibration method may require a little preparation work for combining the data sets, the equating process is easy since no additional transformation needs to be made: the IRT analysis already places all items on the same scale. It should be noted, however, that item parameters produced from the joint calibration utilise both data sets. This may or may not be desirable. As we mentioned in previous sections, there may be good reasons for retaining the item parameters from a particular test, in which case the anchor or shift method may be better.

Common Person Equating Method

In some situations, administered tests need to be released publicly. Under these circumstances, no items from a previous test can be used as common items for future tests because the items are already in the public domain. A number of methods can be used for equating such tests. First, a common person equating method can be used, where both tests are taken by a group of respondents. Thus the relative difficulties of the items from two tests can be established when the same group of people have taken both tests. Figure 12.3 shows the combined data diagrammatically.

When common person equating is undertaken, some considerations should be given to the selection of the respondents taking both tests. In some cases, the

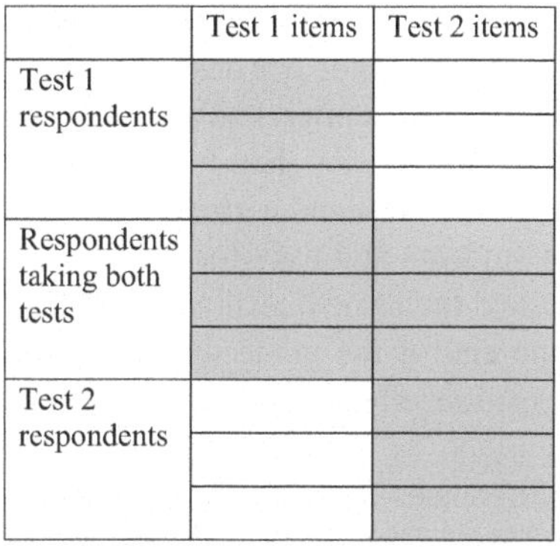

Fig. 12.3 Data structure for common person equating

so-called "off-shore" equating has been used where respondents in another geological location are asked to take both tests. In other cases, students from a different grade level for which the tests are targeted have been used as samples for equating. In all these cases, care must be taken to ensure the sample selected for taking both tests have the same student background characteristics as the respondents taking individual tests. For example, if the students in the equating sample have schooling with a different curriculum, then the relative item difficulties may be different from those for students taking individual tests. These will threaten the accuracies of the equating process.

Further, when tests are administered to the equating sample of students, the item position effect should be considered. For example, if test 2 is always placed at the end of test 1, then test 2 items will appear to be more difficult than they actually are. One might consider the possibility of administering the two tests on separate days to ensure that students are fresh when taking each test. In general, the equating study should be carried under similar circumstances as the original tests, both in terms of the selection of students for equating, and the test administration conditions. This can be a challenge sometimes.

A variation to the common person equating method is to develop some "secure" items (not released items) for equating purposes. A subsample of students taking test 1 will also take the secure items. Similarly, a subsample of students taking test 2 will also take the same secure items. Thus the two tests can be linked through the secure items.

Overall, there are many variations to equating methods. But the most important consideration is to maintain item invariance in equating studies.

Horizontal and Vertical Equating

The terms *horizontal* and *vertical* equating have been used to refer to equating tests aimed for the same target level of students (horizontal) and for different target levels of students (vertical). For example, if a number of tests for grade 4 students are administered, the equating of these tests is often known as horizontal equating. On the other hand, if a test for grade 4 students and a test for grade 6 students are equated, it is often known as vertical equating. There are many more challenges for vertical equating than for horizontal equating. For example, common items between a grade 4 test and a grade 6 test will likely be items that are more difficult for grade 4 students and easy for grade 6 students so that students from both grades can be asked the same questions. As a result, frequently these common items are placed at the end of the grade 4 test and at the beginning of the grade 6 test, so that item position effect, as discussed previously, will be an issue in equating the tests. Further, as the curricula for grade 4 and grade 6 students are different, there are differences in terms of "opportunity to learn" between students from two different grades. That is, a low performance of a grade 4 student on an item may not be related to the low ability of the student, but the low performance may be attributable to grade 4 students not having the opportunity to learn the topic being tested.

The challenges for vertical equating are especially pertinent for measures of growth over time. For these reasons, some assessment programs do not make an attempt to link between performances across different grade levels.

Equating Errors (Link Errors)

In the same way that standard errors are computed for estimated item parameters and population parameters (such as mean ability scores of groups of respondents), equating processes also contribute to the uncertainty of estimated statistics. Frequently, the margin of error associated with the equating process, termed equating error here (also known as link errors), has been ignored in many empirical studies. In fact, equating errors are quite substantial, and they should be reported, particularly when comparisons between tests are made (e.g., trends over time).

In the OECD PISA assessment, the equating error is computed in a relatively simple way (see, e.g., OECD 2009). The idea is to capture the variability of estimated item parameters in two different tests. We use our example data set in Table 12.2 to illustrate the computation of equating error. Column 4 (headed "Difference") shows the differences between estimated item difficulty parameters in two tests, when the two tests are aligned on the same scale. A standard error statistic of these differences is computed. That is, the equating error is computed as the standard deviation of these differences divided by the square root of the number of items. (Note that in PISA 2006 and 2009, owing to the unit structure of the items, some adjustments to the equating error are made. See PISA technical reports). The equating error computed in this way reflects the amount of variation in item parameters across two tests. If common items have identical estimated item difficulties across two tests, then the equating error will be zero. In contrast, if items vary in their difficulties from one test to another, then the equating error will be large. The equating error is an indication of the amount of uncertainty caused by the sampling of items (for equating), similar to the uncertainty caused by the sampling of students. Mathematically, the equating error can be expressed as

$$\text{equating error} = \frac{\text{standard deviation of}\left(\delta_i - \delta_i'\right)}{\sqrt{L}} = \frac{\sqrt{\frac{\sum_{i=1}^{L}\left(\delta_i - \delta_i'\right)^2}{L-1}}}{\sqrt{L}} \tag{12.6}$$

where L is the number of link items, δ_i is item parameter for item i in test 1, and δ_i' is the item parameter for item i in test 2 that has been aligned with test 1 (i.e., the average item difficulties for both tests are equal so that $\sum_{i=1}^{L}\left(\delta_i - \delta_i'\right) = 0$).

The magnitudes of equating errors are generally large (in comparison with trend or growth estimates). In our example, if the three outlier items are removed in Table 12.2, and the remaining 29 items are used as common items for equating, the equating error is 0.038 logit. This translates to about 4 PISA score points, which is about the magnitude of, or slightly larger than, the standard errors of country mean scores in PISA. The equating errors in PISA vary between domains and between years of comparison, but the magnitude is also around 4 PISA score points.

Additional Notes and References Michaelides and Haertel (2014) use a bootstrap method to estimate the equating error due to the sampling of common items. They note that such equating error will not become smaller as respondent sample size increases, so this error can become the dominant source of variability in assessment results.

Wu (2010) discusses the magnitude of measurement, sampling and equating errors in large-scale assessments and draws cautions over these errors.

More generally, sources of equating error can also come from the sampling of respondents. Kolen and Brennan (2004) discuss in details the methods for estimating the standard errors of equating due to the sampling of examinees.

How Are Equating Errors Incorporated in the Results of Assessment?

Equating errors are *systematic* in the sense that if the shift constant is incorrectly estimated, we may over- or under-estimate the ability and item difficulties by a constant amount. If ε is the error in the equating shift constant, then every respondent's ability is incorrectly estimated by ε. If comparisons are made between groups of respondents in one test, for example, between girls and boys in the 2013 test in our example, then all boys and girls will have the same equating error of ε in their ability estimates, so that the difference between the mean scores between girls and boys will not contain the ε term as it is cancelled out in the subtraction between the mean scores of the two groups. Consequently, it has been argued that for comparisons between groups within one test, equating errors should not be incorporated. On the other hand, for comparisons between two different tests where equating has been carried out, equating errors should be taken into account. The technical reports of OECD PISA explain the treatment of equating errors in some details (see, for example, OECD 2009).

Challenges in Test Equating

Test equating is complex, not just because of technical aspects of test equating, but also because of real-life challenges of keeping test items invariant. For example, on-going curriculum reforms occurring around the world will change emphases in student learning, so that items will not remain the same in terms of difficulties, and tests will not remain the same in terms of content balance. There is always a tension between writing a test to reflect current student learning, and writing a test to maintain the historic test construct for monitoring trends. In some cases, these tensions cannot be reconciled. In the National Assessment of Educational Progress (NAEP) program in the United States, for example, the test for monitoring trends is separate from the test for monitoring current student performance.

For equating tests administered at the same time, there are also many challenges. Item position effect has been discussed in previous sections. There have also been suggestions that the sequencing of test items has an effect on the item difficulties. For example, in the first cycle of OECD PISA, reading items always appeared before mathematics and science items in a test booklet, as reading was the major domain. This could have an impact on the calibrated difficulties of the mathematics and science items. In PISA, even when test items have been rotated to appear in different positions in a number of test booklets, a significant *booklet effect* was still found that needed to be taken into account (Adams and Wu 2002). In PISA, test booklets were randomly distributed to students within a classroom, so it was expected that the average student abilities for different test booklets should be the same as the sample size of respondents was large. However, it was found that the average abilities of students taking different test booklets were not the same, and the pattern of differences between booklets was consistent across most countries, indicating that there were systematic biases in the estimation of student abilities.

Summary

This chapter provides some rationales for the need to equate tests. A number of equating methods are introduced to illustrate equating procedures. These methods include common item equating and common person equating. Among common item equating, the shift method, anchoring method and joint calibrations method are discussed. These methods focus mostly on the equating of the location of the ability scale, and not on the scale factor of the ability scale. In this chapter, we have not considered traditional, non-IRT equating methods, such as Equipercentile equating (see Holland and Rubin 1982; Kolen and Brennan 2004).

This chapter highlights a number of challenges to test equating. If item response data fit the IRT model perfectly, we only need very few common items between two tests to equate the tests. But in real-life, item response data never fit the mathematical model. Factors such as differential item functioning and item position effect have a significant impact on the reliability of equating results. In fact, we cannot stress enough that many common items (well in excess of 30) are needed to equate two tests. Equating errors come from both the sampling of students and sampling of items. Increasing the sample size of students will reduce equating errors from the sampling of students. However, there is a limit to the number of common items for equating owing to limits on test length, so equating error from the selection of common items will likely remain large. Careful planning at the test design stage is essential to mitigate the effects of model violations on equating.

It should be noted that there is no single best method for equating. In practice, we often try out different equating methods and compare the results. If the results from different equating methods vary a great deal, we try to understand why there are such differences. For example, are the differences caused by a few problematic items? Are there plausible explanations for why some items have invariance issues? Does the IRT model matter in the equating: is a 2PL or 3PL model better than the Rasch model, or vice versa? Which equating method produced most credible results, and why? These kinds of investigations will help us improve future assessment designs. Even in well-established large-scale international studies, equating methods are not firmly set. For example, PISA used different equating methods for different cycles. To date, equating remains one of the more difficult technical challenges in assessments.

Discussion Points

(1) Discuss the relative merits of the shift, anchoring and joint calibration methods of equating. What criteria would you devise in order to choose among these three equating methods?

(2) Discuss how the presence of equating error would affect the results from a linking study between two tests with a set of common items. What are some practical steps during the instrument design or test preparation stage that can minimize potential equating error later on?

Exercises

Q1. The following table shows estimated item parameters for 20 link items from two tests separately calibrated. If Test 2 needs to be placed on Test 1 scale, compute the equating transformation using the shift method. Also compute the equating error

Link item	Test 1	Test 2
1	0.66	0.98
2	−0.64	−0.21
3	0.80	1.20
4	−1.44	−1.08
5	−0.59	−0.44
6	−0.86	−0.46
7	1.41	1.91
8	−0.74	−0.59
9	−1.41	−1.26
10	−0.15	0.22
11	−1.11	−0.84
12	−0.69	−0.34
13	−0.85	−0.55
14	−0.40	−0.15
15	0.64	1.10
16	0.24	0.49
17	0.17	0.46
18	−1.07	−0.85
19	−0.15	0.31
20	−0.73	−0.67

References

Adams R, Wu M (eds) (2002) PISA 2000 technical report. PISA, OECD Publishing

Holland PW, Rubin DB (eds) (1982) Test equating. Academic, New York

Kolen MJ, Brennan RL (2004) Test equating, scaling, and linking: methods and practices, 2nd edn. Springer, New York

Michaelides MP, Haertel EH (2014) Selection of common items as an unrecognized source of variability in test equating: a bootstrap approximation assuming random sampling of common items. Appl Measur Educ 27(1):46–57

OECD (2009) PISA 2006 technical report. PISA, OECD Publishing

Wu M (2010) Measurement, sampling and equating errors in large-scale assessments. Educ Meas Issues Pract 29(4):15–27

Chapter 13
Facets Models

Introduction

In previous chapters, the item response models specify a probability function for the chance of being successful on an item, or for obtaining partial/full credit on an item. These probability functions depend on the values of person ability and item difficulty. That is, if a person is of high ability, and/or if the item is easy, the probability of success will be high. On the other hand, if the ability is low and/or the item is difficult, the probability of success will be lower. In short, the probability of success is a function of ability and item difficulty.

There are situations where factors other than ability and item difficulty will also have an impact on students' scores on an item. For example, if markers are employed to mark extended response items, markers may vary in their leniency/harshness. Consequently, two students with the same ability and similar item responses may have different scores on the same item because two different markers mark the students' work, and one marker is more lenient than the other marker. In this case, to model the probability of test scores, we need to take into account of ability, item difficulty and marker harshness. A simple way to model marker harshness in the probability function is shown in Eq. (13.1).

$$p = P(X = 1) = \frac{\exp(\theta_n - (\delta_i + \rho_m))}{1 + \exp(\theta_n - (\delta_i + \rho_m))} \tag{13.1}$$

where θ_n is the ability of person n, δ_i is the item difficulty of item i, and ρ_m is the harshness/leniency of marker m. Comparing Eq. (13.1) with Eq. (7.1), it can be seen that the marker harshness is added to the item difficulty, so the effect of a harsh marker is to make an item more difficulty (i.e., harder to obtain a higher score), and the effect of a lenient marker is to make an item easier (i.e., easier to obtain a higher score). We note that the model in Eq. (13.1) assumes that a marker has the same harshness/leniency across all items. If a maker is sometimes lenient and sometimes

M. Wu et al., *Educational Measurement for Applied Researchers*,
DOI 10.1007/978-981-10-3302-5_13

harsh depending on the item, then we will need to model an interaction term γ_{im} to be added to the item difficulty component. γ_{im} is the adjustment that needs to be made to the item difficulty (δ_i) and the overall marker harshness (ρ_m) of marker m, as shown in Eq. (13.2).

$$p = P(X = 1) = \frac{\exp(\theta_n - (\delta_i + \rho_m + \gamma_{im}))}{1 + \exp(\theta_n - (\delta_i + \rho_m + \gamma_{im}))} \quad (13.2)$$

The terms added to the item difficulty (δ_i) in Eqs. (13.1) and (13.2) are sometimes known as facets, to indicate different factors that have an influence on the item difficulty. Many factors can have an influence on the difficulty of an item, or, more generally, on the probability of success on an item. For example, test administration mode, such as computer-delivered tests or paper-and-pen tests, may have different difficulty levels even if the test items are the same. In that case, we may model a test–delivery–mode parameter and add it to the item difficulty. So test delivery mode is a facet that impacts on students' probabilities of success.

DIF Can Be Analysed Using a Facets Model

We note that one of the DIF methods discussed in Chap. 11 (IRT method 2) is a facets model. In Eq. (11.2), the DIF parameter, D_{gi}, is a facet term that is added to the item difficulty, δ_i, in much the same way as ρ_m is added to δ_i in Eq. (13.1).

An Example Analysis of Marker Harshness

An example data set is used to demonstrate the analysis of marker harshness using a facets model. The data set contains the ratings of four markers who marked 428 students' project work. Each student's project is marked on four criteria (labelled as item 1 to item 4 in the data set) by four markers. For each item, a partial credit scoring from 0 to 4 is used, where 0 is the lowest score and 4 is the maximum score for an item. Table 13.1 shows an excerpt of the data set.

This data set is a complete marking design whereby every marker marked all students and all items. In many cases, the marking design will be incomplete as it will generally be too expensive to have all makers mark all students' work. Typically, each marker will only mark a subset of students' work, but there are links across markers so that the results from all markers can be placed on the same scale.

A facets model with interaction terms for a partial credit model [extension of Eq. (13.2)] is used to fit the markers' data, where the average marker harshness (ρ_m) and an interaction term between marker and item (γ_{im}) are modelled. Specifically, Eq. (13.3) shows the fitted partial credit facets model.

Table 13.1 An excerpt of a data set by four markers on four items with partial credit scoring between 0 and 4

Record number	Student ID	Marker ID	Item 1	Item 2	Item 3	Item 4
1	1	1	3	1	4	1
2	1	2	2	0	2	1
3	1	3	3	0	3	2
4	1	4	2	0	1	2
5	2	1	3	1	4	1
6	2	2	2	1	3	1
7	2	3	2	0	3	3
8	2	4	2	0	1	2
9	3	1	3	2	0	1
10	3	2	2	2	0	1
11	3	3	4	4	3	3
12	3	4	2	3	1	4
…	…	…	…	…	…	…

$$p = P(X = k) = \frac{\exp \sum_{j=1}^{k} \left(\theta_n - \left(\delta_{\bullet} - \tau_j + \rho_m + \gamma_{im}\right)\right)}{D} \tag{13.3}$$

where D is the sum of probabilities across all response categories for the item. $\delta_{\bullet}$ and τ_j parameterisation of the PCM is discussed in Chap. 9.

An R package for IRT analysis, TAM (Kiefer et al. 2012), is used to carry out the analysis. The results of the facets analysis are shown in Table 13.2.

In Table 13.2, i1 to i4 refer to items 1 to 4; rater1 to rater4 refer to the four markers. In addition to item difficulty and rater harshness measures, interaction terms (e.g., i1:rater1) are also estimated. All estimated measures are in logit unit. The standard errors of the estimates are also given. Owing to model identification constraints to the parameters, some standard errors are not available (NA).

A number of observations can be made regarding the results shown in Table 13.2. First, the item difficulty estimates for the four items show that items 2 and 3 are much more difficult than items 1 and 4 (1.366 and 0.906 vs. −0.016 and −0.343). Second, the marker harshness measures show that marker 3 is more harsh (0.302) while marker 1 is more lenient (−0.433). A positive rater estimate means that one has to add an amount to the item difficulty, making the item more difficult than for the average rater. A negative estimate means that one has to subtract an amount to the item difficulty, making the item easier than for the average rater. The item-by-rater interaction terms show that markers are not consistently harsh or lenient across all items. For example, for item 1, marker 1 is even more lenient than his/her average leniency, as a further 0.454 has to be subtracted from the item difficulty estimate. On the other hand, marker 3 who is a harsher marker than other markers is even more harsh on item 1 (add a further 0.396 to item 1 difficulty).

Table 13.2 Results of marker harshness analysis

Parameter	Facet	Parameter estimate	Standard error of estimate
i1	item	−0.016	0.029
i2	item	1.366	0.028
i3	item	0.906	0.023
i4	item	0.343	0.029
rater1	rater	−0.433	0.019
rater2	rater	−0.006	0.019
rater3	rater	0.302	0.019
rater4	rater	0.137	NA
i1:rater1	item:rater	−0.454	0.029
i2:rater1	item:rater	0.273	0.028
i3:rater1	item:rater	0.306	0.026
i4:rater1	item:rater	−0.125	NA
i1:rater2	item:rater	0.027	0.029
i2:rater2	item:rater	−0.237	0.028
i3:rater2	item:rater	0.021	0.026
i4:rater2	item:rater	0.189	NA
i1:rater3	item:rater	0.396	0.029
i2:rater3	item:rater	−0.198	0.029
i3:rater3	item:rater	−0.298	0.026
i4:rater3	item:rater	0.101	NA
i1:rater4	item:rater	0.031	NA
i2:rater4	item:rater	0.162	NA
i3:rater4	item:rater	−0.029	NA
i4:rater4	item:rater	−0.164	NA

In contrast, marker 3 is not so harsh on item 3 (add 0.302 and subtract 0.298, with a net harshness measure of 0.004 on item 3). These results are more easily seen from graphical displays of the expected scores of items for each marker. Figure 13.1 shows these expected scores curves.

Each graph in Fig. 13.1 shows the four markers' expected scores, as a function of ability, on an item. First, note that the curves for items 1 and 4 are generally higher than for items 2 and 3, indicating that items 1 and 4 are relatively easier, with higher expected scores across the ability range as compared to items 2 and 3. This is also reflected in the estimates shown in Table 13.2.

For the expected scores curves, the solid black line is marker 1's curve. Dashed red line is marker 2's curve. Dotted green line is marker 3's curve. Dot-dash blue line is marker 4's curve. It can be seen that marker 1 (solid black line) tends to be more lenient than other markers, with a curve generally higher than for the other markers. On the other hand, marker 3 (dotted green curve) is quite harsh on items 1 and 4, with expected curves below that of others, but marker 3 has an average harshness for items 2 and 3, with expected scores curves in the middle of other curves.

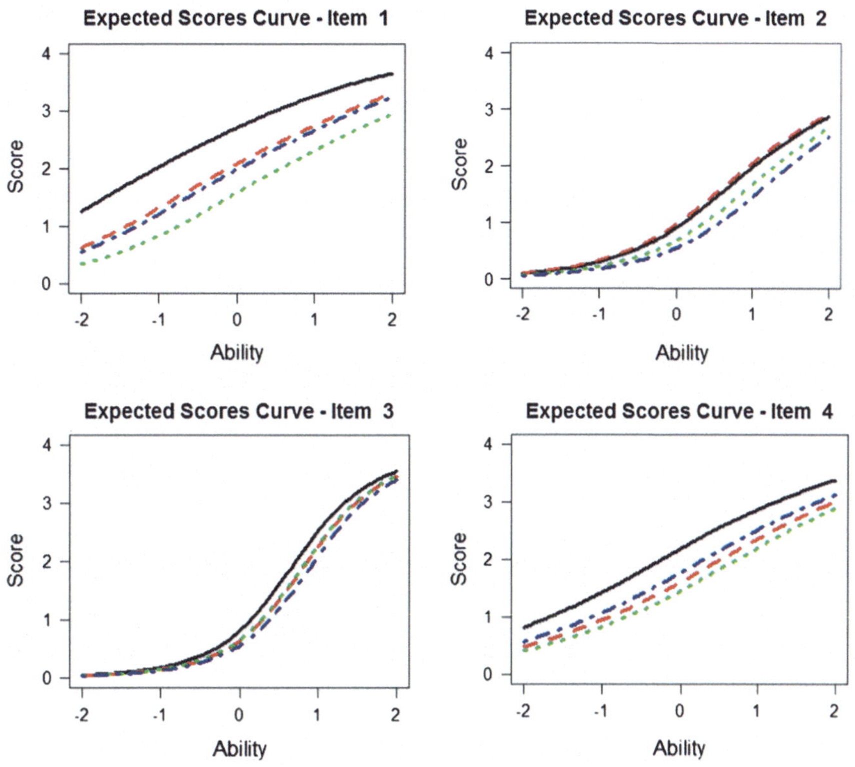

Fig. 13.1 Expected scores curves for four items and four markers

Further, an assessment of marker agreement shows that the agreement is best for item 3, and worst for item 1. For item 3, the four marker curves are relatively close together, with the widest difference of less than half a score point. In contrast, for item 1, the markers vary greatly in their harshness, with a difference of more than one score point across the ability range between the most lenient and most harsh markers.

Further, an examination of the observed scores curve against the expected scores curve can provide information about how the markers apply the marking guide to separate students. Figure 13.2 shows 16 graphs (4 items by 4 markers) of expected scores curve versus observed scores curve.

A few observations can be made regarding Fig. 13.2. Rater 3's observed scores curves tend to be steeper than the expected scores curves, indicating that the ratings given by Rater 3 discriminate students' work more than the ratings of other raters. Typically in assessment, the aim is to separate students by their ability levels, the more one can utilise the full range of scores in the marking guide, the more we can discriminate between low and high levels of students. In this sense, Rater 3 is doing a better job than other raters. Sometimes raters may be *playing safe*, so they avoid giving extreme scores. In that case, students' scores tend to be bunched up in the middle of the scoring range, and we will have less power in separating students by

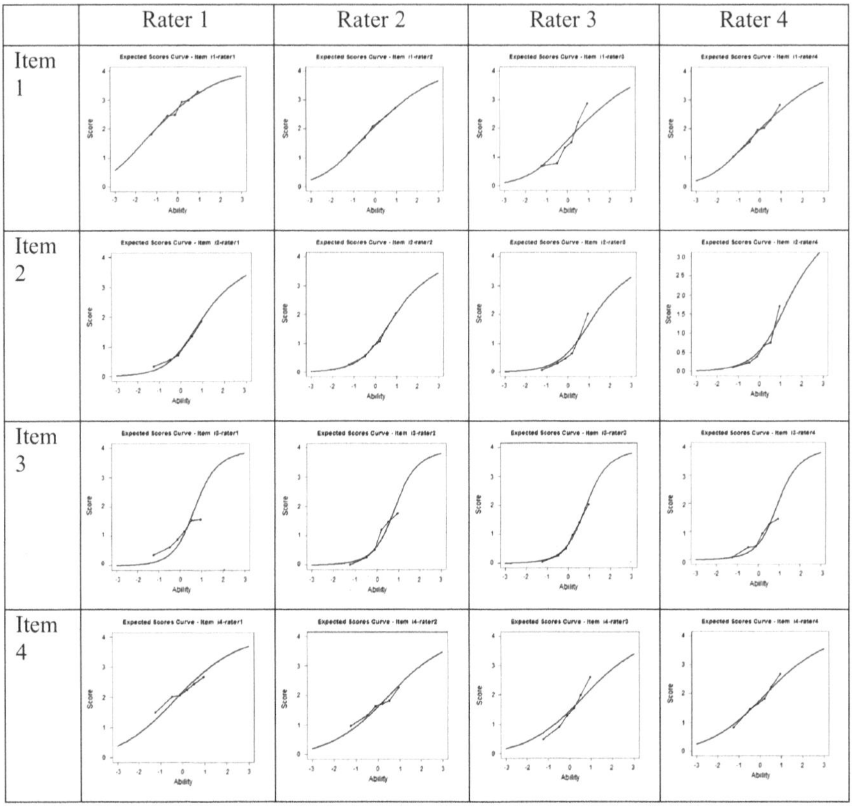

Fig. 13.2 Expected scores versus observed scores curves

their ability levels. In other cases, raters may not be applying the marking guide appropriately so that some high ability students may get lower scores and vice versa, resulting in low discrimination of students. Consequently, in addition to comparing rater harshness, rater discrimination can also be examined to provide some information on how the raters are using the marking guides.

An analysis such as the one presented above can provide a great deal of information about marker behaviour and marker agreement. Such information can be useful feedback to the markers during marker training or for future improvement.

The following are a few notes about the facets models more generally.

Ability Estimates in Facets Models

In Eqs. (13.1) and (13.2), we take note that the model takes into account of the marker harshness in the item difficulty measure when abilities are estimated.

Fig. 13.3 Sampled records from dataset shown in Table 13.1

	id	marker	i1	i2	i3	i4
4	1	4	2	0	1	2
6	2	2	2	1	3	1
8	2	4	2	0	1	2
9	3	1	3	2	0	1
12	3	4	2	3	1	4
13	4	1	3	0	4	1
15	4	3	2	0	3	2
18	5	2	2	2	0	1
20	5	4	2	2	1	3
22	6	2	1	2	0	1
23	6	3	4	4	4	3
26	7	2	2	1	4	2
28	7	4	3	0	1	3

....

That is, if a student is marked by a harsher marker, then the item difficulty is adjusted upward, so that the probability of success is based on the student taking a more difficult item than the probability would be if the student was marked by a lenient marker. In this way, under the facets models of Eqs. (13.1) and (13.2), the ability estimates computed have taken marker harshness into account. That is, the ability estimates adjust for marker harshness. If two students obtain the same overall score but are marked by different markers, their ability estimates may not be the same. In the above example, however, since every student is marked by the same four markers, students with the same score have the same ability estimate. However, if we take a random sample of the data so that there is now an incomplete marking design, then students with the same score may not have the same ability estimates. As an illustration, a random sample of the data set is selected, as shown in Fig. 13.3, where the first column is the record number in the original data set; the second column is student ID; the third column is marker ID, followed by the scores on four items given by the marker.

Using a facets model with an item-by-marker interaction term, the ability estimates for the first seven students are given in Fig. 13.4, where the first column is student ID; the second column is the number of items (e.g., if two markers marked the student's work, there will be 8 items); the third column is the total score the student received on the items; the fourth column is the possible maximum score; the fifth column is the weighted likelihood ability estimate (WLE); and the last column is the standard error of the WLE ability estimate (See Chap. 7 about weighted likelihood ability estimate).

From Figs. 13.3 and 13.4, it can be seen that student 3 and student 7 were each marked by two markers. Markers 1 and 4 marked student 3's work, while markers 2 and 4 marked student 7's work. Both students obtained a raw score of 16 (total score from two markers). But the ability estimate for student 3 is 0.346, and the ability estimate for student 7 is 0.433. This is because marker 2 is a harsher marker than marker 1, so student 7's ability is adjusted slightly higher given that he/she had a harsher marker than student 1 did.

id	N.items	Score	Max	WLE	WLE error
1	4	5.0	16	0.059	0.414
2	8	12.0	32	0.148	0.279
3	8	16.0	32	0.346	0.279
4	8	15.0	32	0.316	0.273
5	8	13.0	32	0.220	0.276
6	8	19.0	32	0.664	0.268
7	8	16.0	32	0.433	0.272
....					

Fig. 13.4 Ability estimates for the first seven students

In summary, one needs to be aware of the adjustment to ability estimates when different facets apply to different students. In the case of markers as a facet term in the IRT model, we would indeed want the ability estimates to be adjusted according to whether students had harsh or lenient markers. But we need to be able to explain to the layperson why students obtaining the same raw score may not have the same ability estimate.

However, sometimes we may have a facet term for which we do not want to use it to adjust for the ability estimates. For example, gender may be used as a facet term, and gender-by-item interaction term can be used to detect DIF in a facets model. Suppose Eq. (13.4) is fitted in an IRT analysis,

$$p = P(X = 1) = \frac{\exp\left(\theta_n - \left(\delta_i + G_g + D_{ig}\right)\right)}{1 + \exp\left(\theta_n - \left(\delta_i + G_g + D_{ig}\right)\right)} \tag{13.4}$$

where g is male or female, so G_g is a gender main effect and D_{ig} is an interaction term between gender group and item. Such an analysis can be used to detect gender DIF through the D_{ig} term, and also to estimate the average difference in abilities between males and females. While G_g and D_{ig} will provide useful information, we need to be aware that the ability estimates produced in this model, θ_n, is an ability where the gender main effect G_g has been taken out. That is, the ability estimates produced are for "gender-neutral" persons, if such persons exist, in much the same way as the removal of marker effect in estimating abilities using Eq. (13.2). In fact, to compute the ability for a male person, we need to subtract G_{male} from θ_n, and to compute the ability for a female person, we need to subtract G_{female} from θ_n. In summary, while we want to remove DIF effect, D_{ig}, from person abilities, but we need to include gender main effect for male or female ability estimates.

Consequently, when a facets model is fitted, you need to be clear about the mathematical model underlying the analysis, so you will know whether the ability estimates are correctly produced. It would also depend on the software program you are using and how a particular program handles the estimation of ability estimates. As a rule of thumb, a factor that is related to items and tests can be modelled as a facet term. A factor that is related to the persons should be treated with care in a facets model. For example, test administration mode (e.g., online versus pen-and-paper) can be a facet as this relates to the tests. Any test administration mode effect should be removed from the ability estimates. In contrast, gender, age group, country of residence and ethnic group are all attributes of persons. That is, a person cannot be without a gender, age, residence or ethnicity, so we will not want to remove any of these factors from a person's ability estimate. In latent regression IRT models (discussed in the latter section of this chapter), person attributes such as gender are often treated as regressors. In general, it will be better to treat person attributes as regressors than as facets. While the effect of a facet can be similarly estimated from using the facets model and using the latent regression model, the estimation of the ability will differ in the two models.

Choosing a Facets Model

As Eq. (13.2) shows, the probability function can be made progressively more complex, to take in all factors that may have an impact on the probability of success of a person. Therefore, there is an inclination to choose a complex model so that all bases are covered. However, there are a number of considerations when we decide on which model to choose. First, a very complex model requires a great deal of data to estimate all the parameters. Using Eq. (13.2) as an example where an item-by-marker interaction term is modelled, it is essentially assumed that there is an item difficulty parameter for each item and marker combination. That is, for item 1 marker 1, the item difficulty is different from the difficulty for item 1 and marker 2. Therefore, to estimate each interaction term well, we need to have sufficient *pieces* of information for a parameter. In our example, marker 1 marked 428 students' work for item 1, so there are 428 *pieces* of information for the estimation of γ_{11} parameter. This seems sufficient. As our example is a complete marking design, there is sufficient information for the estimation of each parameter in Eq. (13.2).

For other data sets, particularly incomplete marking designs where markers may not mark every item or every student, we need to be more cautious about choosing an IRT model. For example, if marker 1 has not marked item 3, then there is no information on γ_{31}, so any software program will have difficulty in estimating this parameter (e.g., we can't estimate girls' average ability if there is no girl in the data set!). Frequently, warning messages from software programs relate to this issue where there is no data to estimate a modelled parameter. As another example, if a test contains multiple choice items and open-ended items where markers only marked the open-ended items, then any item-by-marker term should not include the multiple choice items, as no marker has marked those items.

Consequently, when choosing an IRT model, we need to assess whether there is sufficient information provided by the data to warrant the estimation of parameters in our model.

An Example—Using a Facets Model to Detect Item Position Effect

This is an example demonstrating the use of the facets model, concurrent equating and item position effect as discussed in Chap. 12 on equating.

In many large-scale studies where the aim is to measure proficiency at group levels (e.g., state level or country level) rather than at individual student level, the use of rotated test booklets is common. In order to cover all content areas in a subject domain, many items are developed, but it would be impractical to test every student on every item. Consequently, the items are allocated to a number of different booklets, with appropriate linkages across booklets so the booklets can be equated.

The purpose of this example is to demonstrate item position effect in test booklets. That is, if an item appears at the end of a test booklet, it will generally be more difficult than if it appears at the beginning of a test, most possibly due to fatigue effect. The results from this example also have an implication for common item equating, where different test booklets contain link items, and the link items are in different positions in the test booklets.

Structure of the Data Set

The data set for this example comes from a test where seven rotated test booklets were constructed, using a Balanced Incomplete Block (BIB)[1] test design, as shown in Table 13.3.

C1 to C7 denote 7 different clusters of items, each containing approximately 20 min of testing material. Each student is administered one booklet, and the seven booklets are distributed randomly within each class of students.

Two data sets are used for this example. The first file contains data from all 7 booklets. The second file contains data from booklets 3, 5 and 6 only. If we use data from all 7 test booklets, then each item appears once in each of the three positions (Blocks 1, 2 and 3). If we use data from booklets 3, 5 and 6 only, then the link items are only the C6 items, and these link items appear at the end of booklet 3, in the middle of booklet 5, and at the beginning of booklet 6. Other items in booklets 3, 5 and 6 are unique to each book and they only appear in one position. The aim of this example is to contrast the differences if the test design is balanced (as for the 7 booklets), and if the test design is not balanced (as for booklets 3, 5 and 6 only).

Figure 13.5 shows an excerpt of the data file for booklets 3, 5 and 6, where every line contains data of a student. The booklet number, gender and item responses for a student are recorded. There are 91 items distributed into the 7 item clusters, but

[1]A BIB test design is one where every cluster of items appears in each position once, and every pair of clusters appears together in one test booklet once.

Table 13.3 Assignment of item clusters to booklets

Booklet	Block 1	Block 2	Block 3
1	C1	C2	C4
2	C2	C3	C5
3	C3	C4	C6
4	C4	C5	C7
5	C5	C6	C1
6	C6	C7	C2
7	C7	C1	C3

each student only took a subset (about 40) of the items. It can be seen that there are gaps in the item responses for each student for the items not administered to the students. The dots at the end of each data line indicate that there are more item responses not shown.

Analysis of Booklet Effect Where Test Design Is not Balanced

First, we analyse the data set containing booklets 3, 5 and 6. A facets model is fitted, as shown in Eq. (13.5).

$$p = P(X = 1) = \frac{\exp(\theta_n - (\delta_i + b_m))}{1 + \exp(\theta_n - (\delta_i + b_m))} \tag{13.5}$$

where b_m is known as "booklet effect" for booklet m. When the data containing item responses from three booklets (shown in Fig. 13.5) are calibrated together, we have concurrent or joint calibration (see Chap. 12 on equating tests). That is, the items in all three booklets are equated through link items across the three booklets. Item parameters for items in cluster C6 are based on the item responses to C6 in all three booklets. If an item has the same difficulty irrespective of which booklet the item appears in, then the b_m terms should be close to zero, since δ_i already represent the item difficulty, there should not be any other adjustment. However, if an item has different difficulties in three different booklets, we will need to have a term b_m to make this adjustment.

Note that we have not included an interaction term between item and booklet. Apart from the items in cluster C6, other items only appear in one booklet. For example, we cannot have an interaction term involving booklet 5 and items in C3, since there is no item response data for this combination. It will be difficult to estimate a parameter without any data for the parameter, in the same way that we cannot estimate the harshness of a marker when the marker has not marked any student's work. Further, the "booklet" term b_m is a constant adjustment for all items in the same booklet. If every item has a different adjustment, then we are essentially assuming that the link items are different when they appear in different booklets. In that case, we have three test booklets containing different items, and different

```
3 F      107     000       930210440413         3430 32214..
3 F      102     010       130111441413         3270 34212..
3 M      114     111       031119440413         3230 31212..
3 F      114     091       131111441413         3239 32292..
3 M      003     000       040300221342         2430 12201..
5 F 01      44411   114        04414431      30   301 ..
5 M 02      44423   141        04414131      30   301 ..
5 M 12      44433   112        14414131      31   711 ..
5 M 03      43144   012        14414132      30   311 ..
.
```

Fig. 13.5 An excerpt of the data file for booklets 3, 5 and 6

students took the three test booklets. Therefore, the booklets would not be linked so there would be an issue with model identification. This is an illustration of deciding on the inclusion of terms in a facets model. A recommendation is to keep the model simple if you can, and ensure that the data can support the estimation of the parameters.

Table 13.4 shows the estimates of the booklet parameters, b_m, in logit units.

The estimated booklet parameters have a standard error of about 0.013, so that "booklet effect" shows statistical significance. The difference between booklet 6 and booklet 3 parameters is 0.347 logit, which is more than half a year of growth. Further, booklet 3 parameter is an adjustment of item difficulties "upwards" (i.e. to make items more difficulty), while booklet 6 parameter is an adjustment downwards (i.e. to make items easier). This is consistent with the fact that the link items (C6 items) are at the end of booklet 3, and at the beginning of booklet 6. The results in Table 13.4 show that the link items do not have the same item difficulties when the items appear in different booklets. A concurrent equating without taking into account of booklet effect will result in incorrect item parameter estimates, leading to incorrect ability estimates. For booklet 3, the items will be assumed to be easier than they actually are, so it will disadvantage students taking booklet 3. In contrast, the items in booklet 6 are assumed to be more difficult than they actually are, so students taking booklet 6 will have over-estimated abilities.

To check this result, percentages correct for each item as it appears in each booklet are computed. Given that the booklets were randomly distributed in each class and there were about 400 students taking each booklet, one would expect the percentages correct for each link item to be similar across the three booklets. Table 13.5 shows some of these items. Reading across each row in Table 13.5, it can be seen that, in general, percentages correct are higher for items appearing in

Table 13.4 Estimated "booklet" parameters in logits

	Estimated booklet parameters, b_m
Booklet 3	0.168
Booklet 5	0.012
Booklet 6	−0.179

Table 13.5 Percentages correct for items in different booklets

Item	Booklet 3 (% correct, question no.)	Booklet 5 (% correct, question no.)	Booklet 6 (% correct, question no.)
29	24% (Q31)	26% (Q19)	30% (Q8)
30	86% (Q29)	89% (Q17)	90% (Q6)
31	77% (Q30)	80% (Q18)	86% (Q7)
32	63% (Q32)	72% (Q20)	71% (Q9)

earlier parts of a test than in latter parts of a test. This is consistent with the estimated booklet parameters in the facets analysis.

Analysis of Booklet Effect—Balanced Design

If a test design is balanced, such as the one shown in Table 13.3 where all seven booklets are used, then the unfairness caused by one link cluster of items at the end of a booklet is offset by one link cluster of items that appears at the beginning of the booklet, so overall, when all items are calibrated concurrently, we would expect a smaller booklet effect.

Table 13.6 shows that the booklet parameter estimates are relatively smaller when all seven booklets are used, as compared to when only three booklets are used, showing that the bias caused by the position of one cluster of items is offset by another cluster of items in the booklet to some extent.

Discussion of the Results

The results from the above analyses highlight the importance of item position effect, whether for common item equating or for rotated test booklets where there is an implicit equating across the booklets. Very often, for vertical equating,[2] common items are embedded in lower grade (e.g. grade 3) and higher grade (e.g. grade 5) tests. Because the common items need to be at appropriate levels for both grades, the common items tend to be difficult for the lower grade students, and easy for higher grade students. It is often the practice to place more difficult items at the end of a test, and easier items at the beginning of a test. If the common items are placed in this way, then item position effect may cause a sizable equating error.

For rotated booklets, it is important to have a balanced test design. While this does not entirely eliminate booklet effect, it helps to reduce it.

[2]We use the term *vertical equating* here to refer to equating tests between different grade levels, e.g., between grades 3 and 5.

Table 13.6 Estimated booklet parameters in logits for all seven booklets

	Estimated booklet parameters, b_m
Booklet 1	0.020
Booklet 2	−0.052
Booklet 3	0.028
Booklet 4	−0.049
Booklet 5	0.100
Booklet 6	0.024
Booklet 7	−0.069

Summary

This chapter explains the facets models and provides examples to illustrate the use of facets models. Facets models are IRT models where there are factors other than item difficulty and person ability influencing the probabilities of success on an item. Differential item functioning is a special case of a facets model. Facets models are useful for estimating marker harshness, as markers also influence a student's score, over and above the difficulty of an item and the ability of the student. An example is presented to demonstrate the use of the facets model to estimate item position effects. In particular, cautions are drawn for common item equating where link items appear in different positions in a test.

It is noted that in a facets model the ability estimates take into account of the facets terms. In the case of the modelling of marker harshness as a facet term, student abilities have been adjusted for marker harshness.

Some guidelines are given regarding the inclusion of facets terms in a model. The main consideration is a check of the amount of data available for the estimation of modelled parameters.

Discussion Points

(1) Many factors other than ability and item difficulty can affect the chance of success of a person on an item. Some of these factors are suitable to model as facet terms, some are suitable to model as latent regression terms. Discuss which factors are suitable as facets and which factors are suitable as latent regression conditioning variables.

(2) How does a "balanced design" of item rotation help in calibrating item difficulties and person abilities? For example, for a student taking booklet 1, cluster 4 items will likely be more difficult than their calibrated values as they appear at the end of the test. How does a balanced test design alleviate item position effects?

(3) In Tables 13.4 and 13.6, we note that a constraint is placed on the booklet parameters so that the sum of the booklet parameter is zero. This is similar to the estimation of DIF (Chap. 11) where the net DIF across all items is zero. This relates to model identification. Discuss why this model identification is needed. What will happen if there is no constraint on the booklet parameters?

Exercises

Q1. Indicate whether you agree or disagree with each of the following statements

In a facets model where marker harshness is modelled, a marker has an estimated measure of −0.8. This indicates the marker is harsher than the average marker	Agree/disagree
In a test analysis where marker harshness is modelled, two students have the same score on the test. The abilities of the two students may differ if they are marked by different markers	Agree/disagree
In a vertical equating between grade 3 and grade 5 students, common items are placed at the end of the grade 3 test and at the beginning of the grade 5 test. Grade 3 students' abilities are likely to be under-estimated	Agree/disagree
If gender is placed in a facets model as a facet term, the ability estimates produced may not reflect the abilities of the students	Agree/disagree
The facets model shown in Eq. (13.3) models a marker harshness term that is constant across the ability range. That is, a marker is equally harsh (or lenient) for high ability and low ability students	Agree/disagree

Reference

Kiefer T, Robitzsch A, Wu M (2012) TAM (Test analysis modules)—an R package [computer software]

Further Reading

Linacre JM (1989) Many-faceted rasch measurement. MESA Press, Chicago

Chapter 14
Bayesian IRT Models (MML Estimation)

Introduction

In this chapter, we introduce a family of IRT models where there is an assumption about the shape of the population distribution of abilities.

When a population distribution assumption is made about the latent trait, the model is often known as a Bayesian IRT model. Such a model is often estimated using the MML (marginal maximum likelihood) estimation methods, as opposed to JML (joint maximum likelihood) or the CML (conditional maximum likelihood) methods where no population distribution assumption is made. Readers are referred to Baker and Kim (2004) about estimation methods for details. While MML refers to an estimation method, the term is often used as if it is a kind of IRT model. The distinguishing feature between MML and, say, JML, is that there is a distribution assumption with respect to the population in MML. MML becomes an alias for the more correctly termed Bayesian IRT models. We note that there are also other estimation methods for Bayesian IRT models.

There are advantages and disadvantages in making an assumption about the population distribution of abilities. The disadvantage is that any assumption made in a model needs to be validated. If an assumption is incorrect, then the results are invalid. On the other hand, when assumptions are valid, the results can be greatly enhanced. Let us use an example to illustrate this concept. In measuring the heights of a randomly selected sample of people, if the distribution of heights in a population is normally distributed, then, given the mean and variance of the heights for a simple random sample, we can make inferences about proportions of people in the population with heights within certain ranges. Without an assumption about the shape of the population distribution, we will not be able to make many inferences about the population beyond the sample of data we collected. Further, sometimes the data may distort the shape of the population distribution. For example, if an ability distribution is normal, but a test is extremely easy so there is a strong ceiling effect with most students obtaining perfect scores, then the sample ability

M. Wu et al., *Educational Measurement for Applied Researchers*,
DOI 10.1007/978-981-10-3302-5_14

distribution will be very skewed, not reflecting the true shape of the population. Having an assumption that the population of abilities is normally distributed will result in a better estimate of the shape of the population distribution.

Consequently, Bayesian IRT models are useful when our focus is on making inferences about a population beyond the sample data we collected. For example, in assessing students' educational outcomes in large-scale surveys where samples of students are selected, Bayesian IRT models will help us make better inferences about the population characteristics of students, provided, of course, that our population assumption is correct.

Bayesian Approach

The term "Bayesian" is coined after Thomas Bayes, an English statistician in the 18th century. For our purposes of explaining Bayesian IRT models, it suffices to say that Bayesian probabilities often involve prior probabilities where the estimation of parameters of interest are based on combining new information collected with the prior probabilities. We provide an example to illustrate this.

Consider a group of basketball players and their hypothetical proficiency distribution in shooting goals. From past experience, we have information of the proficiency distribution as shown in Table 14.1. The first column shows the long-term rate of success (success rate when many attempts are made) where 0.1 indicates a success rate of 1 goal in 10 attempts, etc. The second column shows the proportion of players with the success rate in column 1. So 80% of the players have a long-term success rate of 0.1. The last column shows the number of players with each success rate if there are 1000 players in total. We have only specified four possible success rates to simplify the example. One can add more success rates in real-life situations for the probability distribution but the process is the same.

Graphically, the probability distribution of shooting proficiency is shown in Fig. 14.1.

The probability distribution shown in Table 14.1 and Fig. 14.1 is known as a prior distribution—information we already have from the past.

Suppose a player, Zack, comes along and attempts to shoot goals. He tries 10 times and is successful on three attempts. That is, Zack's success rate on this occasion is 3 out of 10. If we are asked to estimate Zack's goal shooting success rate, we may arrive at 0.3 as an estimate, based on Zack's performance on this

Table 14.1 Hypothetical probability distribution of basketball shooting success rates

Rate of success in shooting goals	Proportion of players	Number of players out of 1000
0.1	0.80	800
0.3	0.15	150
0.5	0.04	40
0.7	0.01	10

Fig. 14.1 Histogram of basketball shooting rate (Prior distribution)

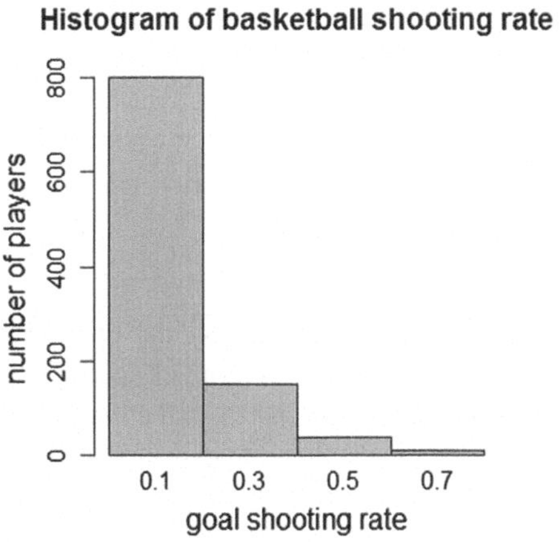

occasion alone and not taking any information of the prior distribution of player proficiencies.

However, using a Bayesian approach, we would proceed as follows. Zack is from the population of players with the prior distribution of shooting rates shown in Table 14.1. If Zack does not attempt to shoot any goals and we have no information about how Zack may perform, and we are asked to estimate Zack's shooting success rate, we will probably choose 0.1, since most players (80%) in the population have this success rate, based on the prior distribution. That is, if a player is randomly selected from the population with no performance information about the player and we are making a guess of the player's success rate, we will be right 80% of the time if we choose 0.1 as the success rate.

But Zack now provides us with more information by making 10 attempts to shoot goals. On this occasion Zack scores 3 out of 10. Based on this new piece of information, we have the opportunity to revise our estimate of 0.1 to take into account of the additional information. The revision of our estimate is made as follows.

The question we ask is "how many players with each long-term success rate in the prior distribution are likely to score 3 out of 10?" We use the success rate of 0.1 as an illustration. If a player's overall success rate is 0.1, then the probability of this player scoring 0, 1, …, 10 goals out of 10 attempts can be computed using a binomial distribution:

$$\Pr(X = k) = \binom{10}{k}(0.1)^k(1 - 0.1)^{10-k} \tag{14.1}$$

where k is the number of goals shot, out of 10 attempts. These probabilities are computed and shown in Table 14.2.

Table 14.2 shows the binomial probabilities of obtaining a score k out of 10, given the long-term success rate is 0.1. Since there are 800 players with success rate

Table 14.2 Probability and expected number of players out of 800 with long-term success rate of 0.1 scoring k goals out of 10

k	0	1	2	**3**	4	5	6	7	8	9	10
Pr(k)	0.35	0.39	0.19	**0.06**	0.01	0.00	0.00	0.00	0.00	0.00	0.00
Expected no of players out of 800	279	310	155	**46**	9	1	0	0	0	0	0

of 0.1, we multiply 800 by the probabilities to obtain the expected numbers of players with success rate of 0.1 scoring k goals out of 10. That is, even though the players' success rate is 0.1 in the long run, out of 10 attempts the players may score higher (or lower) than the expected score of 1. In particular, we note that 46 players out of 800 are expected to score 3 out of 10. We repeat this process for players with success rates of 0.3, 0.5 and 0.7, and find the expected number of players to score k out of 10 given a particular success rate. The results are summarized in Table 14.3.

Getting back to estimating Zack's proficiency given that he scored 3 out of 10, we look through the column with the heading $k = 3$ in Table 14.3. We find that in the population, 46 players with success rate of 0.1 are expected to score 3; but we also note that 40 players with success rate of 0.3 are expected to score 3 goals, 5 players with success rate of 0.5 and 0 players with success rate of 0.7 are also expected to score 3 goals out of 10. We can express these numbers proportionally in Table 14.4.

From Table 14.4, we can see that, if a player scores 3 out of 10, there is a 0.51 chance the player has a long-term success rate of 0.1; 0.44 chance of a success rate of 0.3, and 0.05 chance of a success rate of 0.5. Therefore, if we are to estimate a

Table 14.3 Expected number of players scoring k out of 10, given a success rate

k	0	1	2	**3**	4	5	6	7	8	9	10
Success rate = 0.1	279	310	155	**46**	9	1	0	0	0	0	0
Success rate = 0.3	4	18	35	**40**	30	15	6	1	0	0	0
Success rate = 0.5	0	0	2	**5**	8	10	8	5	2	0	0
Success rate = 0.7	0	0	0	**0**	0	1	2	3	2	1	0

Table 14.4 Expected number and proportion of players with each success rate to score 3 out of 10

$k = 3$	Expected number of players to score 3 out of 10	Proportion of players out of 91 (posterior probabilities)
Success rate = 0.1	46	0.51
Success rate = 0.3	40	0.44
Success rate = 0.5	5	0.05
Success rate = 0.7	0	0.00
Total	91	1.00

player's long-term success rate when the player scores 3 out of 10, we will choose 0.1 or 0.3, as these two are the most likely to result in the observation of 3 goals out of 10. In particular, a success rate of 0.1 is more likely than a success rate of 0.3. Column 3 in Table 14.4 is known as the posterior distribution, derived after combining new information (a score of 3 out of 10) with the prior information (how many players are in each success rate group, Table 14.1).

Mathematically, we express the posterior distribution as $\Pr(\text{long}-\text{term success rate}|\text{score k out of } 10)$. The symbol "|" within the expression denotes "given the fact that" in the context of conditional probability. So this is the probability of a particular long-term success rate given that a player scored k out of 10 in a tryout.

The probabilities calculated in Table 14.2 can be expressed as $\Pr(\text{score } k \text{ out of } 10|\text{long}-\text{term success rate})$, the probability of scoring k out of 10 in one tryout, given a particular long-term success rate. This is somewhat opposite to the posterior distribution.

The prior distribution can be expressed simply as $\Pr(\text{long}-\text{term success rate})$, the probability of a particular long-term success rate.

All the above somewhat complex description of the process of Bayesian approach can be elegantly summarised using Bayesian mathematics. Following Bayes theorem on conditional probabilities,

$$\Pr(A|B) = \frac{\Pr(B|A)\Pr(A)}{\Pr(B)} \tag{14.2}$$

where A and B denote two different events. The posterior probability can be expressed in terms of the prior and marginal probabilities as follows.

Let $\Pr(A|B)$ be the posterior distribution so A is the long-term success rate and B is the event of scoring k out of 10 attempts. Then, using Bayes theorem Eq. (14.2), the posterior distribution can be expressed as

$$\begin{aligned}&\Pr(\text{long}-\text{term success rate}|\text{score k out of } 10)\\ &= \frac{\Pr(\text{score } k \text{ out of } 10|\text{long}-\text{term success rate})\Pr(\text{long}-\text{term success rate})}{\Pr(\text{score } k \text{ out of } 10)}\end{aligned} \tag{14.3}$$

where the denominator is the sum of the numerator term over all success rates. That is, the denominator term can be expressed as follows:

$$\begin{aligned}&\Pr(\text{score } k \text{ out of } 10)\\ &= \sum_{\text{success rates}} \Pr(\text{score } k \text{ out of } 10|\text{long}-\text{term success rate})\Pr(\text{long}-\text{term success rate})\end{aligned}$$

As an illustration, for long-term success rate of 0.1 in the above example, the posterior probability is computed as

$$\begin{aligned}
&\Pr(\text{success rate} = 0.1|\text{score 3 out of 10})\\
&= \frac{\Pr(\text{score 3 out of 10}|\text{success rate} = 0.1)\Pr(\text{success rate} = 0.1)}{\Pr(\text{score 3 out of10})}\\
&= \frac{0.06 \times 0.8}{0.06 \times 0.8 + 0.27 \times 0.15 + 0.12 \times 0.04 + 0.01 \times 0.01}\\
&= 0.51
\end{aligned}$$

Some Observations

For the example of Zack's performance, we note the difference in the conclusions that might be drawn between a non-Bayesian approach and Bayesian approach. In a non-Bayesian approach, the estimate of Zack's long-term success rate will be 0.3, as he scored 3 out of 10. In the Bayesian approach, it's a toss-up between 0.1 and 0.3, with 0.1 as a slightly more probable success rate for Zack. The reason for this discrepancy is that in the Bayesian approach, the prior distribution is taken into account as well as current information. Since the majority of players have a success rate of 0.1, the estimated success rate for Zack is being "pull" towards this part of the population. In general, the prior distribution can be regarded as weights when estimates are computed. For the dense part of the prior distribution, the weights are large.

Second, the Bayesian approach produces a probability distribution (the posterior distribution) instead of a point estimate for estimating individual proficiency. Under the Bayesian approach, the inference about Zack's success rate is made in terms of probabilities, rather than a point estimate. We make statements such as "there is 0.51 chance that Zack has a success rate of 0.1, 0.44 chance a success rate of 0.3, 0.05 chance a success rate of 0.5". In other words, Zack's long-term success rate is described by a probability distribution instead of a single number. Of course if a single number is needed we may make rules about how to derive a point estimate from the posterior distribution. But this is almost an addendum to the Bayesian approach rather than part of the approach.

Third, since the prior distribution provides weightings when estimates are computed, if players are from different populations where the prior distributions are different, the derived posterior distribution will be different. For example, if a player is in the NBA (National Basketball Association) and also obtains 3 goals out of 10 attempts, the posterior distribution for the NBA player will be "higher up" than the posterior distribution for Zack, since the prior distribution of NBA players will show higher success rates of goal shooting. Zack may cry unfair treatment given that he obtained the same score as the NBA player, yet Zack's estimated success rate is lower. We might provide an explanation that on this occasion the NBA player was a little unlucky while Zack was quite luck. For people who are not convinced whether prior information should play a part in estimating performance, perhaps we will ask them to answer the question that if 100 attempts are given to Zack and the NBA player next time, which player will they back as the winner?

Unidimensional Bayesian IRT Models (MML Estimation)

Using the Bayesian approach to estimate student assessment outcomes, the process mirrors the basketball example. First, a population distribution of student abilities is assumed to be the prior distribution, typically a normal distribution, although the mean and variance of the prior distribution are unknown parameters to be estimated from the data. Students' performance on a test is the additional information we obtain, as for Zack's trial of 10 shots. The probabilities for this additional part are modelled using IRT models such as the ones discussed in this book. The IRT part and the population part are then combined to form posterior distributions of abilities for each student. More formally, a Bayesian IRT model is presented below for a normal prior distribution and dichotomous Rasch model for the item responses.

Population Model (Prior)

Let θ represent a student's ability and δ represent the difficulty of an item.

$$\theta \sim N\left(\mu, \sigma^2\right)$$

where μ is the mean of the population distribution of abilities and σ^2 is the variance. Both μ and σ^2 are unknown and they are estimated from the item response data. In estimating these two parameters, essentially we ask the question what values of μ and σ^2 are most likely to have the item responses we collected. We use $g(\theta)$ to denote the normal density function.

Item Response Model

$$P(X = 1) = \frac{\exp(\theta - \delta)}{1 + \exp(\theta - \delta)}, \quad P(X = 0) = \frac{1}{1 + \exp(\theta - \delta)}$$

We use $\mathbf{X}$ to denote a vector of item responses (0s and 1s) across all items for a student, and $f(\mathbf{X}|\theta)$ to denote the product of the item response probabilities across all items for a student. So $f(\mathbf{X}|\theta)$ is the probability of observing a response pattern on a set of items in a test given the ability of a respondent, θ.

Putting together the population model and item response model using Bayes theorem, the posterior distribution for each student is given by

$$h(\theta|\mathbf{X}) = \frac{f(\mathbf{X}|\theta)g(\theta)}{\int f(\mathbf{X}|\theta)g(\theta)d\theta} \tag{14.4}$$

where the denominator of Eq. (14.4) is the integral of the numerator over all θ values. We denote the denominator as $f(\mathbf{X})$ which is the probability of a response pattern, $\mathbf{X}$. Formally, $f(\mathbf{X}) = \int f(\mathbf{X}|\theta)g(\theta)d\theta$.

As an exercise, readers can match the elements of Eq. (14.4) with those of Eq. (14.3) to gain an understanding of how the posterior distribution is formed.

In estimating the Bayesian IRT model, we may use a maximum likelihood approach and estimate parameters of the model by maximising the probability of all observed item response patterns, $\prod f(\mathbf{X}_n)$, where the product (denoted by $\prod$) is over all students. We note that under the Bayesian approach, the parameters estimated are δ_i (item difficulties), μ and σ^2. Individual abilities (θ_n) are not parameters of the model since θ has been integrated out in $f(\mathbf{X})$. However, to make inferences about individual students' ability, posterior distributions Eq. (14.4) for each student can be formed. To find a point estimate for each student's ability, a number of methods can be taken. First, the mean of the posterior distribution can be computed as an ability estimate. This estimate is called EAP (expected a posteriori). Alternatively, the mode of the posterior distribution can be used as a point estimate for a student's ability, and this statistic is called the MAP (maximum a posteriori). However, we stress that neither the EAP nor MAP are parameters in a Bayesian IRT model. They are additional statistics computed after the model has been estimated.

Some Simulations

We carry out some simulations to contrast the Bayesian and non-Bayesian approaches. In these simulations, the abilities are generated from a normal distribution with mean 0 and variance 1. The items are dichotomous and item difficulties are evenly distributed from −2 to 2. The Rasch model is used to model the item responses. For the Bayesian approach we use the MML estimation method, and for the non-Bayesian approach we use the JML estimation method.

We note that the JML method estimates individual person's ability and makes no assumptions about the population distribution. So, for checking how well the JML method recovers population distribution parameters such as the mean and variance, a two-stage process is used where individual ability estimates are first computed using JML, and then the mean and variance are computed based on the individual ability estimates. In contrast, in MML estimation method, the mean and variance are parameters in the model so they are directly estimated and not re-constructed from ability estimates.

However, when individual abilities are compared using the JML and MML methods, we use the WLE (weighted likelihood) ability estimates from JML, and EAP estimates after the MML model is estimated.

Simulation 1: 40 Items and 2000 Persons, 500 Replications

Frequently, a test is administered in one class period so a test length of around 40 items is common. The sample size of students for each simulation run is 2000, considered large enough to provide stable item parameter estimates. Five hundred simulation replications are carried out. The data are analysed using JML and MML.

Comparison of Population Characteristics

Table 14.5 shows that MML produces better population parameter estimates than JML. In particular, both JML and MML produce unbiased population mean estimate, but JML overestimates the variance. The reason for the overestimation is because the variance estimate under JML is reconstructed using ability estimates and each individual ability estimate contains measurement error, inflating the variance. The problem of the overestimation of the variance is more severe when the test length is short.

Additional Notes

The estimation of variance using individual ability estimates from JML is said to be attenuated by measurement error so that the observed variance is larger than the true variance. The observed variance can be disattenuated using test reliability. Similar to the relationship in classical test theory

$$\text{variance(true score)} = \text{reliability} \times \text{variance(observed score)}$$

the variance computed using IRT abilities in logits can also be disattenuated as follows

$$\begin{aligned}\text{variance(population abilities)} = {} & \text{reliability} \\ & \times \text{variance(ability estimates from JML)}\end{aligned}$$

This adjustment generally works well when the test difficulty matches student abilities. The adjustment does not work so well when a test is very easy or very difficult for the students.

Table 14.5 Compare JML and MML in recovering population characteristics–40 items

	Generating value	JML	MML
Mean	0	0.00	0.00
Variance	1	1.18	1.00

Comparison of Ability Estimates for Individual Students

Given that the MML estimation takes into account of prior information about the population distribution, one would expect the ability estimate for an individual student under MML to be more accurate since more information is included in the estimation (Consider the example of predicting Zack and the NBA player's long-term success rates in shooting basketball goals). In the simulations, we use root mean squared error (RMSE) to measure how close the estimated ability is from the true ability (generating value). Root mean squared error is computed by calculating the difference between the estimated and the true abilities, square the difference, then average across all replications, and finally take the square root of the average so that the RMSE is in the unit of logit and is interpretable on the logit scale. Five generating abilities are used to make the comparisons between JML and MML where the WLE ability and EAP ability estimates are computed respectively. Table 14.6 shows the results.

The RMSE values in Table 14.6 show that EAP ability estimates from MML method are a little closer to the true ability than the WLE ability estimates from JML, across all ability range. As an aside, we also note that the RMSE is the smallest for abilities closer to the middle of the distribution when the test is better targeted to the respondents. Lastly, we note that even when the test is well targeted, the magnitude of the RMSE is still very large (>0.3 logit), so that a test of 40 score points does NOT measure an individual well.

In addition to RMSE, we also examine the bias in estimating abilities. For each generating ability, we compute the average WLE and average EAP values across 500 replication respectively. Table 14.7 shows the results.

In Table 14.7, the average of WLE estimates across 500 replications is close to the true ability for each of the five generating value, showing no obvious bias. However, for EAP ability estimates under MML, there is a clear bias in the estimated abilities which are being "pulled" towards the middle of the ability distribution. It is sometimes said that the EAP ability estimates are "shrunken" towards

Table 14.6 RMSE for JML and MML methods in recovering individual abilities for 40 items

Generating ability	JML WLE RMSE	MML EAP RMSE
−2	0.48	0.45
−1	0.39	0.35
0	0.36	0.33
1	0.39	0.35
2	0.46	0.46

Table 14.7 Average WLE and EAP ability estimates across 500 replications for 40 items

Generating ability	JML WLE average	MML EAP average
−2	−2.03	−1.73
−1	−1.01	−0.89
0	−0.01	−0.01
1	1.02	0.90
2	2.00	1.71

the mean of the distribution where the distribution is more dense. We have also observed this phenomenon in the example of Zack's basketball shooting rate. Overall, the bias in EAP is more severe for the abilities at the extremes of the ability distribution. But for the majority of students who are close to the mean of the distribution, the bias is not severe.

In summary, EAP ability estimates from MML are biased as seen from the mean estimates in Table 14.7, but for individual students' ability estimates, EAP estimates are, on average, closer to the true ability than WLE ability estimates from JML as seen from the RMSE in Table 14.6. Further, as there are fewer students at the extreme ends of the ability distribution, the larger bias of EAP estimates at these extreme ability values does not contribute too much to the overall RMSE.

Simulation 2: 12 Items and 2000 Persons, 500 Replication

In this simulation, only 12 item responses are generated for each student. This test length is chosen to reflect the rotated test design in PISA where each student may take around 12 test items for a subject domain, particularly for the minor domains. Five hundred replications are carried out in the simulation. Table 14.8 shows the recovery of population parameters: the mean and variance of the population ability distribution.

Comparison of Population Characteristics

When the test length is short (12 items in this case), the estimation of the population mean is still very good under both the JML and MML methods. However, the population variance is considerably overestimated under the JML method. In contrast, MML still recovers the population variance well even though there are only 12 items (score points) per student.

Table 14.8 Compare JML and MML in recovering population characteristics for 12 items

	Generating value	JML	MML
Mean	0	0.00	0.00
Variance	1	1.60	1.00

Comparison of Ability Estimates for Individual Students

Table 14.9 shows the RMSE of individual student ability estimates from JML and MML methods.

Table 14.9 shows that when the test is short, EAP from MML method has considerably smaller RMSE for abilities in the middle range of the distribution than WLE from JML method. The bias of ability estimates is shown in Table 14.10.

The bias in EAP ability estimates from MML is much more severe than the bias in WLE from JML method (Table 14.10). However, for very low and high abilities where the bias is the worst, not many students are located at these extremes. In contrast, EAP does a good job in recovering the abilities in the middle range of abilities where more students are located.

Summary of Comparisons Between JML and MML Estimation Methods

When one is choosing between the JML and MML estimation methods (non-Bayesian and Bayesian), considerations should be given to the focus of the assessment. If the focus is on population characteristics, such as in international comparative surveys like TIMSS and PISA where there is no interest in measuring individual students accurately, MML clearly is a better choice. The simulations show that even when the test length is very short for each student, the collective statistics such as the mean and variance of ability distributions are still estimated very well under MML.

However, if the main goal of the assessment is to provide individual student abilities, JML provides less biased results particularly for students at the extremes of the ability distribution.

Table 14.9 RMSE for JML and MML methods in recovering individual abilities for 12 items

Generating ability	JML WLE RMSE	MML EAP RMSE
−2	0.79	0.80
−1	0.72	0.57
0	0.67	0.45
1	0.72	0.55
2	0.76	0.80

Table 14.10 Average WLE and EAP ability estimates across 500 replications–12 items

Generating ability	JML WLE average	MML EAP average
−2	−2.11	−1.34
−1	−0.99	−0.66
0	0.01	0.01
1	1.03	0.69
2	2.08	1.32

Plausible Values

In large-scale international studies such as TIMSS and PISA, the term "plausible values (PV)" has been used frequently in the data provided by these studies. There have been confusions about what plausible values are and how they are used. In this section, we provide some explanations about plausible values.

First we remind readers about posterior distributions discussed earlier in this chapter. Column 3 of Table 14.4 shows the posterior probabilities. The posterior probabilities are probabilities of each possible (plausible) success rate of Zack, derived by combining Zack's performance in a "test" and the population distribution of proficiencies (goal shooting rates). Table 14.11 shows the posterior distribution once again.

To describe Zack's success rate, we do not provide a single number but we provide a probability distribution as follows: the likelihood of Zack's success rate of 0.1 is 0.51, a success rate of 0.3 is 0.44, etc. Alternatively, we can "draw" observations from the posterior distribution to represent Zack's success rate:

$$0.3, 0.3, 0.1, 0.3, 0.1, 0.1, 0.1, 0.5, 0.1, \ldots$$

These observations are drawn according to the probabilities in Table 14.11. These observations are known as plausible values. If a frequency graph of the plausible values is plotted, we will have the histogram of the posterior distribution, as shown in Fig. 14.2.

Table 14.11 Posterior distribution of success rates

	Posterior probability
Success rate = 0.1	0.51
Success rate = 0.3	0.44
Success rate = 0.5	0.05
Success rate = 0.7	0.00

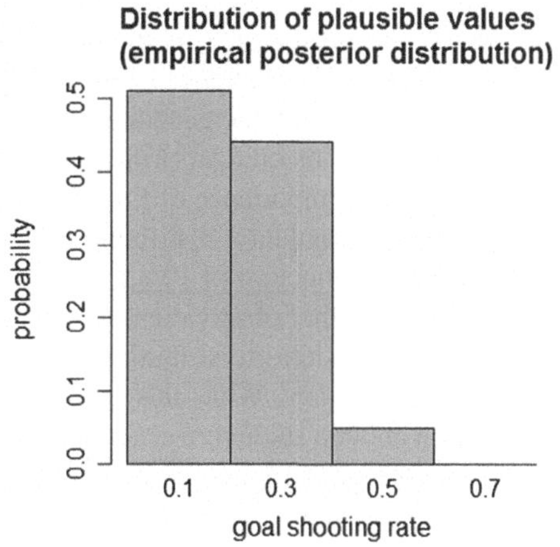

Fig. 14.2 Distribution of plausible values (empirical posterior distribution)

That is, if we are asked the question of “what is Zack’s success rate of basketball goal shooting?”, we can answer as the following: “0.3, 0.3, 0.1, 0.3, 0.1, 0.1, 0.1, 0.5, 0.1, …” are all possible (plausible) values of Zack’s success rate. Similarly, for student achievement measures, plausible values are the likely student achievement measures given the item responses of a student and the prior ability distribution.

Simulation

To illustrate some properties of plausible values, we carry out a simulation for a data set with 2000 persons and 40 items, and we use MML estimation method to fit the item response model. The prior distribution of abilities is assumed to be normal with mean 0 and variance 1. We then draw plausible values for each student. Figure 14.3 shows some results.

The top picture in Fig. 14.3 is a posterior distribution built using PVs for a student whose true ability is −0.88 and his test score is 13 out of 40. The posterior distribution formed by PV shows that this student’s ability is most likely to be around −1, with some likelihood to be in the range of −2 to 0.

The middle picture in Fig. 14.3 shows the distribution built by the PVs for all the students. This distribution is actually the (empirical) prior distribution. Mathematically, we can express the collection of PVs across all students (so across all response patterns) as

$$\begin{aligned} \int h(\theta|\mathbf{X})f(\mathbf{X})d\mathbf{X} &= \int \frac{f(\mathbf{X}|\theta)g(\theta)}{f(\mathbf{X})} f(\mathbf{X})d\mathbf{X} \\ &= g(\theta)\int f(\mathbf{X}|\theta)d\mathbf{X} \\ &= g(\theta) \end{aligned} \tag{14.5}$$

where **X** is a vector of response pattern, and the integration is over all response patterns. We recall that $g(\theta)$ is the prior distribution of students’ abilities.

As a check, in the bottom picture of Fig. 14.3, we have plotted the theoretical prior distribution using the generating abilities in the simulation. It can be seen that the empirical and theoretical prior distributions (middle and bottom pictures) are very similar. The importance of Eq. (14.5) is that the distribution formed by PVs is the (estimated) population distribution. So inferences about the population can be made from the collection of PVs.

We make a further observation by comparing the top and bottom distributions in Fig. 14.3. The posterior distribution for a student is considerably “narrower” than the prior distribution. What this means is that when there is no item response information about a student, we can only make a guess of the student’s ability using the prior distribution. There is much uncertainty in this guess. But when we have additional information about the performance of a student on a test, we make a

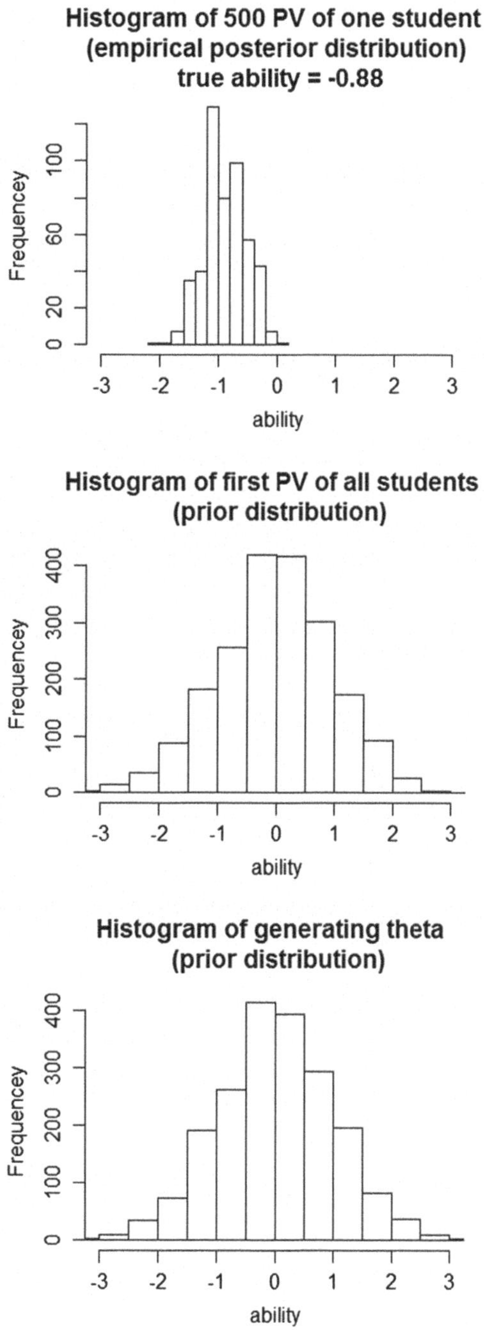

Fig. 14.3 Posterior and Prior distributions

guess of a student's ability using the posterior distribution which has much less uncertainty. In fact, the "width" (or variance) of the posterior distribution is directly related to the length of the test. If a test is very long, then the posterior distribution will be quite narrow giving us more precise locations of the student's ability. On the other hand, when a test is short, the posterior distribution will be wider so we are less certain about where a student is located. In fact, the variance of the posterior distribution is a measure of measurement error, and PV's have been used as a means to incorporate measurement errors into the computations of standard errors.

Use of Plausible Values

The fact that the collection of PVs across all students forms the population distribution allows us to use PVs to make inferences about the population ability distribution. But one may ask the question why this is necessary, given that an assumption is made about the prior (typically a normal distribution), and the mean and variance of this prior are directly estimated from the item response data. So we should already have an estimate of the population distribution without drawing plausible values. There are several reasons for using plausible values in large-scale surveys such as TIMSS and PISA.

First, the process of "scaling" (estimation of the IRT model) is complex, as many student background variables are also incorporated in the IRT model. The inclusion of student background variables is through latent regression (discussed in the next section). The estimation of such complex IRT model requires specialist software and expertise. It is envisaged that secondary data analysts may not have access to such specialist software, so the provision of plausible values is to allow data analysts to use common statistical packages to analyse the data. That is, plausible values can be used to form ability distributions and inferences can be drawn about the distributions without further IRT scaling. Therefore data analysts without special training in IRT can carry out standard statistical analyses using plausible values.

The second reason for providing plausible values is for the computation of standard errors of statistics. For example, for simple random samples, the standard error for the population mean is $\frac{\sigma}{\sqrt{n}}$, where n is the sample size and σ is the standard deviation. But this estimate of standard error is not appropriate under complex sampling, where the standard errors are not easily derived. In large-scale surveys, cluster and stratified sampling is often used for practical considerations in test administration and for increasing the efficiencies of the samples (reducing sampling errors). The standard errors of statistics are typically computed using a replication method (see, for example, technical reports of TIMSS and PISA (e.g. OECD 2009a)). In replication methods, student ability estimates are repeatedly sampled to obtain variations in the statistics of interest in order to estimate standard errors. Since a collection of plausible values across all students represent the population distribution (i.e., the prior distribution), plausible values can be sampled in replication methods for computing standard errors. For details, see PISA data analysis manual (OECD 2009b).

In relation to the use of plausible values, we should stress that PVs are not suitable as individual students' ability measures to be provided to students. This is because PVs are random draws from the posterior distributions. Two students with the same response pattern are likely to have quite different PVs as measurement error is typically large. However, PVs are used as individual ability measures in the contexts of forming aggregate statistics for a population.

Latent Regression

In discussing the facets model in Chap. 13, latent regression has been mentioned in relation to factors influencing the probability of success. In the facets model, *facets* refer to factors (other than person ability and item difficulty) that have an impact on the probability of success on an item. One can sometimes regard factors such as gender, grade, ethnicity as facets, even though these factors are person attributes, and not item attributes such as booklet or rater (see Chap. 13). Strictly speaking, person attributes should be modelled as latent regression variables. In this section, we will explain about the latent regression model and its relationship with the facets model.

Facets and Latent Regression Models

The Bayesian IRT model has two parts: a population part and an item response part. The population part assumes that the abilities come from a population distribution, and typically the unidimensional model is assumed to be a normal distribution with mean μ and variance σ^2, where μ and σ^2 are estimated together with item parameters. For the simplest model, the population distribution can be written as

$$\theta \sim N\left(\mu, \sigma^2\right) \tag{14.6}$$

where θ denotes person ability parameter.

The simplest item response model (dichotomous Rasch model) can be written as

$$\Pr(X = 1) = \frac{\exp(\theta - \delta)}{1 + \exp(\theta - \delta)} \tag{14.7}$$

where δ denotes an item difficulty parameter.

When there is a facet term, Eq. (14.7) can be written as

$$\Pr(X = 1) = \frac{\exp(\theta - (g_r + \delta))}{1 + \exp(\theta - (g_r + \delta))} \tag{14.8}$$

where g denotes one facet term (e.g., raters), and g_r denotes one level (or category) of the facet (e.g., one rater). That is, the item difficulty can be thought of as being altered by an amount that is equal to g_r (e.g., a harsher rater makes the item more difficult for a respondent as the score for the respondent is likely to be lower).

In contrast, for the latent regression model, attributes relating to persons (e.g., gender, SES, grade, motivation) may be identified as possibly having an impact on the latent trait, θ. So the population model can be written as

$$\theta \sim N\left(\mu + \alpha x + \beta y + \ldots, \sigma^2\right) \tag{14.9}$$

That is, the mean of the ability distribution where a person comes from is determined by a set of factors, x, y, …. These factors are known as regressors because the expression "$\mu + \alpha x + \beta y + \cdots$" resembles part of a regression model. For example, x could be 1 or 0 depending on the gender group; y could be a SES measure of the person (in which case, the regressor may be a continuous (rather than categorical) variable). Notice that x, y, …, etc. are known values for each person. The regression coefficients, α, β, … are to be estimated, together with μ and σ^2.

For example, consider the case of one regressor, namely, gender. The population model can be written as

$$\theta \sim N\left(\mu + \alpha g, \sigma^2\right) \tag{14.10}$$

where g takes the value 0 for a boy and 1 for a girl. That is, the ability distribution for boys is

$$\theta \sim N\left(\mu, \sigma^2\right) \tag{14.11}$$

and the ability distribution for girls is

$$\theta \sim N\left(\mu + \alpha, \sigma^2\right) \tag{14.12}$$

As a result, there are two prior distributions depending on the membership of a respondent to a group.

If a latent regression model is fitted, the estimate of α can be regarded as the average difference in mean abilities between girls and boys.

Relationship Between Latent Regression Model and Facets Model

If θ is the ability of a girl, then, from Eq. (14.12),

$$(\theta - \alpha) \sim N(\mu, \sigma^2) \tag{14.13}$$

So it is possible to write the following item response model for girls,

$$\begin{aligned} \Pr(X = 1) &= \frac{\exp((\theta - \alpha) - \delta)}{1 + \exp((\theta - \alpha) - \delta)} \\ &= \frac{\exp(\theta - (\alpha + \delta))}{1 + \exp(\theta - (\alpha + \delta))} \end{aligned} \tag{14.14}$$

For boys, the item response model is simply

$$\Pr(X = 1) = \frac{\exp(\theta - \delta)}{1 + \exp(\theta - \delta)} \tag{14.15}$$

If we let $g_1 = \alpha$, and $g_0 = 0$, then, combining Eq. (14.14) and Eq. (14.15), the item response model can be written as

$$\Pr(X = 1) = \frac{\exp((\theta - g_r) - \delta)}{1 + \exp((\theta - g_r) - \delta)} \tag{14.16}$$

It can be seen that Eq. (14.16) looks identical to Eq. (14.8). However, if a facets model is run, it will be assumed that θ comes from a normal distribution with mean μ and variance σ^2 for *both* boys and girls. This, of course, is incorrect. So a population model mis-specification has occurred.

Two observations can be noted about using facets model or latent regression model for a variable such as gender.

First, the average difference between mean abilities of girls and boys can be estimated in both models. In the facets model, the difference is $g_1 - g_0$. In the latent regression model, the difference is α. These two estimates should be the same (up to the accuracy of the estimations).

Second, if ability estimates are computed, the facets model will produce incorrect abilities, as the population model for both boys and girls are assumed to come from a common normal distribution. The reason is that, in the facets model, the facet is considered as a factor contributing to the item difficulty, so when ability is estimated, this source of item difficulty is removed (or adjusted for). For example, rater harshness is adjusted for when ability is computed. In the case of raters, it is desirable to estimate ability when there is no rater effect. But in the case of gender, we do not want to know the ability of a person when the person has a *neutral* gender, or no gender. In contrast, in the latent regression model, abilities are computed with the correct population models where the mean for girls is $\mu + \alpha$, and μ for boys, so the gender effect is *incorporated* in the estimation of abilities.

Summary

This chapter explains the Bayesian IRT models typically estimated using the marginal maximum likelihood (MML) method. Bayesian IRT models contain two parts: population part and item response part. The population distribution is known as the prior. For an individual student, the estimated ability is not provided as a point estimate, but expressed as a probability distribution known as the posterior. Point estimates of abilities can be computed after the model has been estimated. Typically, EAP is used as a point estimate for ability under MML. Comparing Bayesian and non-Bayesian models, it is shown that Bayesian models provide better population estimates such as the mean and variance of the ability distribution. In contrast, non-Bayesian models provide less biased individual ability estimates. Plausible values are random draws from individual students' posterior distributions. They are frequently provided in large-scale surveys for secondary data analysts to explore the data using standard statistical software packages. Latent regression relates to the specification of multiple prior distributions. When factors influencing the probability of success on an item relate to person attributes, these factors should be modelled as regressors in latent regression models.

Discussion Points

(1) From prior data, it has been found that drivers under 40 years-old have considerable higher accident rates than drivers over 40 years-old. Driver A is 25 and Driver B is 45 years-old. Both drivers have not had an accident in the past 3 years. An insurance company sets premium rates based on age as well as individual driver's record of accidents over a three-year period. If a Bayesian approach is used by the insurance company, would Drivers A and B pay the same premium? Discuss this from the point of view of the insurance company, and also from the point of view of the drivers.

(2) The choice of an IRT model depends on the purposes of an assessment. Surveys such as PISA and TIMSS mainly focus on the performance of a country. In contrast, state-wide testing programs are often interested in measuring individual students. Discuss the relative merits between Bayesian and non-Bayesian IRT models in relation to the purposes of an assessment.

(3) Discuss which ability estimates (WLE, EAP, plausible values) are most suitable for representing the abilities of individuals.

(4) Discuss why plausible values are useful in large scaled educational assessment studies.

Exercises

Q1. Indicate whether you agree or disagree with each of the following statements

Two basketball players both scored 5/10 in goal shooting. Player A comes from the local high school while Player B comes from the state basketball team. If we use a Bayesian approach to estimate the long-term goal-shooting rates, Player B will have a higher estimated rate	Agree/disagree
In a Bayesian IRT model, the prior is a normal distribution with mean 0. If a student's WLE ability estimate under JML (non-Bayesian) is 0.8, the student's EAP ability estimate will be less than 0.8	Agree/disagree
EAP ability estimates are more accurate than WLE ability estimates in that the bias in the EAP estimates is smaller	Agree/disagree
The variance of a posterior distribution for an individual student will be smaller if the test length is longer	Agree/disagree
2000 plausible values for a student are generated. The mean of these plausible values can be used as the EAP estimate for the student	Agree/disagree
When plausible values across all students are collected, we have the prior distribution	Agree/disagree
The variable "test-delivery mode (computer versus paper)" should be regarded as a facet term and not as a regressor term	Agree/disagree
The variable "grade (years in school)" should be regarded as a facet term and not as a regressor term	Agree/disagree

References

Adams RJ, Wilson M, Wu M (1997) Multilevel item response models: an approach to errors in variables regression. J Educ Behav Stat 22:47–76

Baker FB, Kim S-H (2004) Item response theory: parameter estimation techniques, 2nd edn. Marcel Dekker, New York

Mislevy RJ, Beaton AE, Kaplan B, Sheehan KM (1992) Estimating population characteristics from sparse matrix samples of item response. J Educ Meas 29:133–161

OECD (2009a). PISA 2009 Technical report. PISA, OECD Publishing. http://dx.doi.org/10.1787/9789264167872-en

OECD (2009b) PISA data analysis manual. OECD publishing, PISA

Wu M (2005) The role of plausible values in large-scale surveys. Stud Educ Eval 31:114–128

Further Reading

Bock RD, Aitkin M (1981) Marginal maximum likelihood estimation of item parameters: application of an EM algorithm. Psychometrika 46(4):443–459

Fox (2010) covers a comprehensive discussion of Bayesian item response modelling with an emphasis on sampling-based estimation methods such as MCMC (Markov chain Monte Carlo)

Fox J-P (2010) Bayesian item response modelling: theory and applications. Springer, New York

For marginal maximum likelihood estimation methods, there are many important papers including Bock & Aitken (1981)

For plausible values, see Mislevy, Beaton, Kaplan and Sheehan (1992) for an application to NAEP (National Assessment of Educational Progress) data. For a non-technical explanation of plausible values, see Wu (2005)

For a more technical discussion on latent regression IRT models, see Mislevy and Sheehan (1989) as well as Adams, Wilson and Wu (1997)

Mislevy RJ, Sheehan KM (1989) The role of collateral information about examinees in item parameter estimation. Psychometrika 54:661–679

Chapter 15
Multidimensional IRT Models

Introduction

The incorporation of a population model discussed in Chap. 14 leads to an extension where the population distribution is a multivariate distribution rather than a univariate one. That is, instead of having a population distribution assumption for a latent ability, θ, there is now a multivariate distribution assumption for a vector of θ for each respondent:

$$\boldsymbol{\theta} = \begin{bmatrix} \theta_1 \\ \theta_2 \\ \cdots \\ \theta_D \end{bmatrix}$$

For example, for each respondent we could be measuring multiple abilities such as mathematics and reading. The distribution for the vector of $\boldsymbol{\theta}$ is typically a multivariate normal distribution with a vector of means $\boldsymbol{\mu}$ and covariance matrix $\boldsymbol{\Sigma}$. That is, it is assumed that the abilities $(\theta_1 \quad \theta_2 \quad \cdots \quad \theta_D)$ are correlated. In IRT terminology, such an IRT model is termed multidimensional item response model (MIRM). If the abilities are not correlated, then we can just use a unidimensional model to scale each ability separately.

In real-life, students' performances in different subject domains are often correlated. For example, a high performing student is often good at mathematics as well as at reading. In PISA 2009, the (latent) correlations between reading, mathematics and science are given in Table 15.1 (OECD 2012, p. 194).

From Table 15.1, it can be seen that the correlations are rather high, but not as high as one, giving support to the use of multidimensional IRT models.

In the simplest case of MIRM, the items are "loaded" on separate dimensions of ability. That is, each item reflects only one ability. For example, if a test measures mathematics and reading abilities, then some items test only mathematics ability

M. Wu et al., *Educational Measurement for Applied Researchers*,
DOI 10.1007/978-981-10-3302-5_15

Table 15.1 PISA 2009 latent correlations between subject domains for OECD countries

	Reading	Science
Mathematics	0.82	0.88
Reading		0.87

and some items test only reading ability. In this case, the term "between-item dimensionality" is used to describe the multidimensional nature of items. In contrast, if an item tests multiple abilities, such as a problem solving item tapping into both reading and mathematics abilities, then we have "within-item dimensionality". In all analyses shown in this chapter, between-item MIRM are used (see Adams et al. 1997).

In the following sections, we explain how multidimensional item response models can be used to assist with the analysis and reporting of student achievement results. In particular, we provide an assessment of the benefits and the limitations of using MIRM, and the circumstances under which the increased complexity of the methodologies adds value to the results obtained. We also contrast the multidimensional models with unidimensional models.

Using Collateral Information to Enhance Measurement

One motivation for using the multidimensional item response model (and student background variables in latent regression) may be illustrated in simple terms through an analogy in physical measurements. Suppose we are interested in predicting how tall a child will be when he/she reaches adult age, we can collect information such as the current height of the child in relation to children of the same age, the heights of the parents, the region the child is from, the gender of the child, etc. With multiple pieces of information, we have a better chance of making a good prediction than with only one or two pieces of information. Similarly, if we know a child is doing well in mathematics, and that the child comes from a high socio-economic family, we may predict that the child should be doing well in reading as well. If we have no information about a child, then it will be difficult to make a prediction about the child's achievement in reading. In the case where we have a reading test score for the child, then the question is whether the additional information on the child's performance in mathematics and the child's socio-economic background can further enhance our measure of the child's reading level, given that there are measurement errors associated with the reading test score. This justification for using collateral information, whether it is student background information or student's other academic results, stems from the aim to improve the measurement of individual student's level of achievement. This may not be the case if the goal of the assessment is to obtain group level measures and not necessarily individual measures. There are methodologies that can directly estimate group level variables without first computing individual student estimates. Under these circumstances, the use of collateral information plays a somewhat different role in the analysis process. We examine the effect of using MIRM and collateral information on individual student estimates and on group level estimates separately.

A Simple Case of Two Correlated Latent Variables

Suppose two tests are given to a student. Each test consists of 10 items. The two tests measure two latent traits that are correlated. This means that if a student is located high on one latent trait, he/she is likely to be located high on the other latent trait. If the correlation is one, then the two tests measure the same latent trait. If the correlation is zero, then the two tests measure two completely unrelated latent traits. In the case of a correlation of one, one should be able to combine the results from the two tests, and provide a single result that reflects a test of 20 items instead of 10 items. If the correlation is zero, then one cannot combine the two test results in any way, and separate reporting on each test is necessary. What if the correlation is 0.8? Can the result on test 1 (as a measure of the first latent variable) be *improved* by drawing upon the result from test 2 so that the reported measure of latent variable one reflects a test of more than 10 items, but not quite 20 items? The short answer is that multidimensional item response model provides us with such a methodology for combining information from different tests according to how well the latent variables are correlated. If two tests assess two latent variables that have a correlation of zero, the multidimensional item response model will simply use the information from each test alone and ignore information from the other test. On the other hand, if the correlation is high, then the multidimensional item response model will use information from both tests in providing estimates of levels on the latent traits.

A simulation is conducted where abilities are generated from a standard normal distribution and item responses are generated using the simple Rasch model. Two sets of 10 item responses are generated based on the same ability for each student, reflecting a situation where two latent variables being assessed have a correlation of one. Figure 15.1 shows a plot of the generating ability and the estimated EAP (expected a posteriori) ability using only the first set of 10 items. Figure 15.2 shows a plot of the generating ability and the estimated EAP (expected a posteriori) ability for the first latent variable after fitting a two-dimensional item response model. It can be seen that, when only the first set of 10 items are used, the estimated ability departs from the generating ability by quite a margin as indicated by the width of the spread in the diagram. In contrast, when a two-dimensional IRT model is fitted, the estimated ability for the first dimension draws upon information from the item responses in the second dimension, and is closer to the generating ability as compared to the unidimensional case.

As the correlation between the two latent variables decreases, the item responses from one test will have less impact on the ability estimate from the other test. If reading, mathematics and science tests were all conducted, then the estimation of

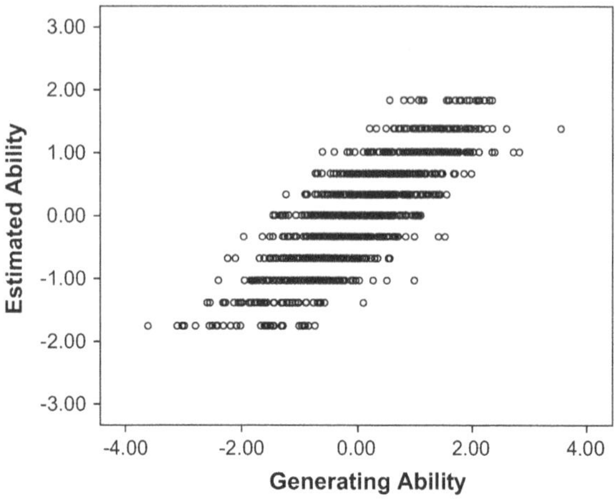

Fig. 15.1 Estimated ability versus generating ability for Dimension 1—fitted unidimensional IRT (10 items on dimension 1)

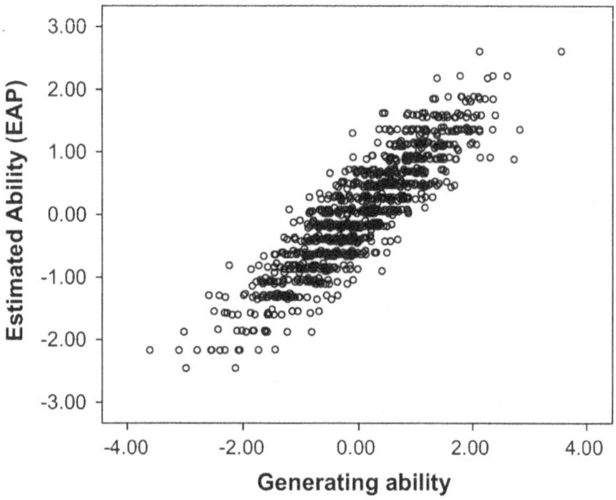

Fig. 15.2 Estimated ability versus generating ability for Dimension 1—fitted 2-D MIRM (10 items on each dimension, correlation = 1 between two dimensions)

ability in each subject area can draw upon information from the other two subject areas, in much the same way as increasing the test length in each test. This is not to say that, if you want accurate measures of reading ability, you should measure mathematics and science abilities. Certainly, the best way to improve precision of reading ability is to increase the reading test length. However, given that mathematics and science tests were administered already and the data are available, there can only be gains by including these data. It should be noted that, in PISA for example, the aim is not to report on individual student results. Rather, group level results are of interest. Therefore, the main reason for carrying out multidimensional item response modelling in PISA is not motivated by the desire to improve individual student ability estimates. The reasons for using multidimensional item response model are discussed in latter sections of this chapter.

A number of observations need to be made in relation to the discussions presented so far. First, if the test length of one test is already quite long (say, more than 50 items), then, the reduction in measurement error from using a multidimensional item response model will not be as significant as when the test length is short. That is, the gain from using a multidimensional item response model may need to be weighed against the complexity of scaling multidimensional data. In addition, when group level measures are of interest, the sample size of students will have a much higher impact on the precision of the population mean estimates than increased test length.

The second observation is that, while on average, the precision of measurement at individual student level will improve by using a multidimensional model, there is a bias in the estimates, caused by the *pull* of information from other dimensions (as well as a pull towards the denser part of the population distribution, as discussed in Chap. 14). For example, if, by chance, a student performed not quite well in a mathematics test than his/her expected performance, then the student's reading score will be affected under the multidimensional approach. Or, if the actual abilities of a student on the two dimensions are far apart as compared to what the model predicts using estimated cohort correlation, then a concurrent two dimensional analysis will result in two ability estimates that are closer together than they would have been if two unidimensional analyses were carried out. This is because in a multidimensional analysis the estimated ability on one dimension can be treated as a weighted average of abilities from all dimensions, so the abilities resulted from a multidimensional analysis tend to be *pulled* towards each other (depending on the magnitude of the correlation), causing a bias in ability estimates. The weights in combining the abilities from all dimensions are related to the strengths of the correlations between the dimensions.

A simulation for 1000 students is conducted where the true abilities are drawn from a bi-variate normal distribution with mean 0, variance 1 for the marginal ability distributions and a correlation of 0.8 between the two abilities. Item responses to 10 items on each dimension are generated. One hundred replications are run. For each student, the same generating ability is used across the replications, but different item responses are generated between the replications. Table 15.2 shows results for the first three students in the simulated data set to illustrate some comparisons.

Table 15.2 Comparison of generating ability and estimated abilities using unidimensional and multidimensional item response model—correlation = 0.8, 100 replications

Student number	Generating ability (correlation = 0.8)		EAP (unidimensional) (averaged over 100 replications)		EAP (multidimensional) (averaged over 100 replications)	
	Dim 1	Dim 2	Dim 1	Dim 2	Dim 1	Dim 2
1	−0.81	1.27	−0.48	0.86	−0.11	0.47
2	0.57	1.13	0.41	0.72	0.64	0.79
3	−1.04	−0.50	−0.74	−0.43	−0.59	−0.47
…	…	…	…	…	…	…
Average RMSE over all students			0.576	0.585	0.525	0.528

Table 15.3 Comparison of estimated population means using unidimensional and multidimensional models—averaged over 500 replications

Generating population mean		Estimated population **mean** and empirical standard error (unidimensional model)		Estimated population **mean** and empirical standard error (multidimensional model)	
Dimension 1	Dimension 2	Dimension 1	Dimension 2	Dimension 1	Dimension 2
0	0	0.00 (se = 0.025)	0.00 (se = 0.026)	0.00 (se = 0.026)	0.00 (se = 0.026)

A number of observations can be made about Table 15.2. Consider a pair of generating abilities for a student. The pair of EAP (unidimensional) estimates tends to be closer in values to each other than the pair of generating abilities is. This is because EAP estimates have a bias (shrunken towards the mean ability), as discussed in Chap. 14, since the multidimensional IRT model is also a Bayesian IRT model with a population distribution assumption. The pair of EAP (multidimensional) estimates tends to be even closer in values than the generating pair. This is because of the pull of estimates between the pair as the mean of each dimension of generating abilities is zero. However, this does not mean that the EAP (unidimensional) estimates will necessarily be closer to the generating values than EAP (multidimensional) estimates will be. For Student 1, it can be seen that EAP (unidimensional) estimates are closer to the generating values than EAP (multidimensional) estimates are. But for Student 2, EAP (multidimensional) estimates are closer to the generating values than EAP (unidimensional) estimates are. A RMSE (root mean squared error) is computed for the ability estimates in Table 15.2. The RMSE is a measure of how close the estimated ability is to the generating ability, on average. A RMSE is computed for each student over the 100 replications, and then an overall average RMSE is computed across the 1000 students. The last row in Table 15.3 shows this average RMSE. It can be seen that EAP (multidimensional) estimates are a little closer to the generating values than EAP (unidimensional) estimates are as indicated by the smaller RMSE values. So, despite the bias, the multidimensional EAP estimates recover the generating abilities better than for the unidimensional model.

Comparison of Population Statistics

In studies such as PISA and TIMSS, the main focus is not on obtaining accurate ability estimates for each student. The main aim is for estimating population statistics. It will be of interest to examine the effect of using unidimensional and multidimensional item response model in estimating population statistics, such as cohort mean scores, the standard errors of the mean scores, the variances and correlations between dimensions.

A simulation is conducted where two abilities are generated for each of 2000 students from a bi-variate normal distribution with mean 0 and variance of 1 for each dimension, and a correlation of 0.8 between the two dimensions. Twenty item responses are generated for each dimension. Each replication re-generates abilities and item responses, so that both sampling error and measurement error are incorporated in the generated data. The estimated population statistics for the two dimensions are recorded after each replication, using both one- and two-dimensional item response models. 500 replications are simulated. The average over the 500 replications is reported in Table 15.3. A standard error is also computed as the empirical standard deviation of the 500 means.

Comparisons of Population Means

Table 15.3 shows a comparison between the Bayesian unidimensional and the multidimensional models.

It can be seen that both the unidimensional and multidimensional models recover the population mean well. In addition, the standard error for the estimated mean is of similar magnitude whether unidimensional or multidimensional model is used.

Additional Notes

Adams (2005) noted that

$$\mathrm{var}(\hat{\mu}) = \frac{\sigma_\theta^2}{NR^E}$$

where σ_θ^2 is the variance of the *true* latent abilities, N is the sample size, and R^E is the reliability of the test (Adams 2005). For a 20-item test in the data sets used in the simulation, the reliability coefficient is around 0.769.

$$\sqrt{\frac{1}{2000 \times 0.769}} = 0.0255$$

This standard error is consistent with those obtained in the simulation as shown in Table 15.3.

Comparisons of Population Variances

Table 15.4 shows a comparison of estimates of population variance using Bayesian unidimensional and multidimensional models. It can also be seen that there is no difference between the unidimensional and the multidimensional models in recovering the population variance.

Table 15.4 Comparison of estimated population variance using unidimensional and multidimensional models—averaged over 500 replications

Generating population variance		Estimated population **variance** and empirical standard error (unidimensional model—Bayesian)		Estimated population **variance** and empirical standard error (multidimensional model)	
Dimension 1	Dimension 2	Dimension 1	Dimension 2	Dimension 1	Dimension 2
1	1	1.00 (se = 0.047)	1.00 (se = 0.047)	1.00 (se = 0.047)	1.00 (se = 0.047)

Comparisons of Population Correlations

Using unidimensional models, the correlation between two dimensions is not part of the models. Therefore, the correlation is computed in a second step using WLE ability estimates after the unidimensional models are estimated. In contrast, in multidimensional models the correlation is directly computed using the estimated variance-covariance matrix. This directly estimated correlation reflects the latent correlation, in contrast to the correlation computed from ability estimates which contain measurement error.

Table 15.5 shows that the two-step approach of computing correlation using abilities from unidimensional models considerably underestimates the correlation, while the multidimensional model recovers well the value of the generating correlation coefficient.

Additional Notes

In much the same way as for the disattenuation of variance estimate discussed in Chap. 14, the correlation computed using ability estimates from unidimensional models can be disattenuated by dividing it by the square root of the reliabilities of the two tests. More specifically, we have:

$$\text{Latent correlation} = \text{correlation of ability estimates}/\text{sqrt(reliability1} * \text{reliability2)}$$

In the simulation example above, to find the latent correlation from the WLE ability estimates, we use

$$\text{latent correlation} = \frac{0.613}{\sqrt{0.769 \times 0.769}} = 0.80$$

where 0.613 is the observed correlation and 0.769 is the reliability of each test.

Table 15.5 Comparison of estimated correlation using unidimensional and multidimensional models—averaged over 500 replications

Generating population correlation	Estimated population **correlation** and empirical standard error (unidimensional model—Bayesian)	Estimated population **correlation** and empirical standard error (multidimensional model)
0.8	0.613 (se = 0.015)	0.799 (se = 0.016)

Comparison of Test Reliability

As the multidimensional model draws information from all dimensions in making inferences about abilities, one would expect that the test reliability to be higher under the multidimensional model. In the simulation example above, the EAP reliability for each dimension is 0.769 when the data for the two dimensions are scaled using unidimensional models separately. However, using the multidimensional model, the EAP reliability is 0.811 for each dimension. The effect of the increase in reliability is equivalent to increasing the test length to about 26 items from 20 items.

Data Sets with Missing Responses

Given that both unidimensional (Bayesian) and multidimensional models recover population mean and variance well, a question arises about the advantages of using multidimensional item response models. One advantage of multidimensional item response model is that, when there are missing item responses, the multidimensional model provides a theoretical underpinning that facilitates the imputation of missing responses, thereby a complete data set can be produced that is easily usable by secondary data analysts.

We use PISA as an example to illustrate the treatment of missing responses. In PISA 2003 there were 13 rotated test booklets, containing test items in reading, mathematics, science and problem solving. Table 15.6 shows the PISA 2003 test design, where M refers to mathematics; R refers to reading; S refers to science and PS refers to problem solving item blocks. Mathematics, being the major domain in PISA 2003, appears in every test booklet. Reading, Science and Problem Solving each appears in 7 of the 13 test booklets. That is, 7 out of every 13 students took reading items, and 6 out of 13 students have missing reading scores. Similarly, for Science and for Problem Solving, $\frac{6}{13}$ of the students do not have scores in that domain. The test booklets are distributed to students at random, so the missing responses are Missing At Random (MAR) as they are missing by design.

Table 15.6 PISA 2003 test design

Booklet	Cluster 1 30 min	Cluster 2 30 min	Cluster 3 30 min	Cluster 4 30 min
1	M1	M2	M4	R1
2	M2	M3	M5	R2
3	M3	M4	M6	PS1
4	M4	M5	M7	PS2
5	M5	M6	S1	M1
6	M6	M7	S2	M2
7	M7	S1	R1	M3
8	S1	S2	R2	M4
9	S2	R1	PS1	M5
10	R1	R2	PS2	M6
11	R2	PS1	M1	M7
12	PS1	PS2	M2	S1
13	PS2	M1	M3	S2

A simulation is carried out to examine the effect of missing item responses when unidimensional and multidimensional item response models are applied. Two abilities are generated for a sample of 1000 students using a bi-variate normal distribution where the correlation is 0.8, and the mean and variance for the marginal distributions are 0 and 1 respectively. Twelve item responses are generated for each of the two dimensions. 25% of the responses on each dimension are then changed into missing values at random, but no student has missing responses on *both* dimensions. That is, 50% of the students have responses on both dimensions; 50% of the students have missing responses in one dimension. The simulation is repeated 100 times.

The results of the simulations show that both the unidimensional and multidimensional models recover the population mean and variance well. However, notable differences are in the correlation estimates and test reliabilities. The estimated correlation between the two latent abilities is 0.53 using WLE ability estimates from unidimensional models, and 0.80 from multidimensional model. Once again the result illustrates that the multidimensional model recovers the correlation much better. Furthermore, we find that the EAP test reliability is 0.5 for each dimension under the unidimensional model, and 0.64 under the multidimensional model.

Production of Data Set for Secondary Data Analysts

To allow secondary data analysts to use the data from a survey, data files containing estimated student scores (e.g., plausible values) are prepared. If a student was not administered items in a subject domain, then, typically, the student's score will be set to missing for that subject domain. This often causes problems for secondary statistical analyses, as many statistical procedures adopt list-wise deletion where the entire case is deleted. In the case of the PISA data sets, as 12 out of every 13 students have missing score(s) in at least one subject area, list-wise deletion will likely remove a substantial amount of data. In PISA, students with missing subject scores have *imputed* scores, so that a complete data set is released. A complete data set is easier to analyse than a data set with missing responses.

Imputation of Missing Scores

The following is an illustration of the idea for the imputation of missing scores. If a student did not sit for a reading test, and no other information is known about the student, then the imputed scores come from the population distribution of reading scores across all students. If the student did sit for the mathematics test, and obtained a high score, say, x, then the imputed reading score will be from the distribution of reading scores of students who obtained x for their mathematics score. Graphically, a bivariate relationship between two scores can be illustrated as shown in Fig. 15.3. It can be seen that the marginal distribution of reading scores (blue curve on the left side of the graph) is the imputation distribution if no information is known about a student. The yellow curve located at 70 on the mathematics scale shows the conditional reading score distribution given that the mathematics score is 70. This conditional distribution has a much narrower spread as compared to the blue curve. Consequently, if the mathematics score is known, then the imputed reading score will be more precise than the imputed reading score

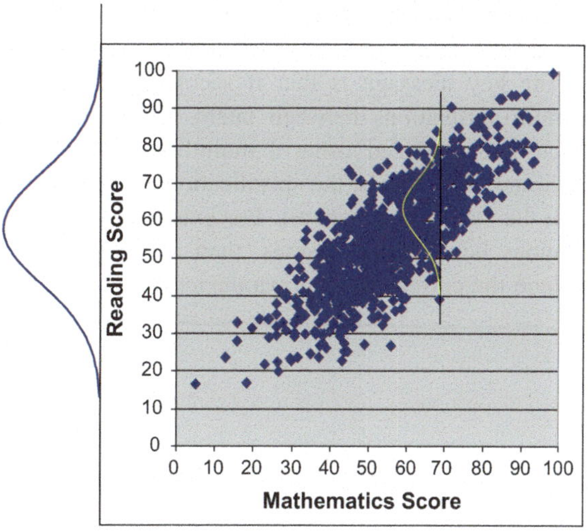

Fig. 15.3 Bivariate relationship between two variables, and marginal distribution of reading scores

when no information is known. Of course, the relationship between mathematics and reading scores has already been established using the observed data. Therefore the imputation of missing values simply uses the parameters of the estimated model which is based on non-missing data. Essentially, the imputation conforms to the estimated model. There is no circularity in this process where the estimation of a model is not affected by imputations.

In the simulation described above where 50% of the students have missing responses in one dimension, unidimensional and multidimensional IRT models are fitted separately to the non-missing data. After the parameters of the IRT models have been obtained, plausible values are generated for all students on both dimensions, including students with missing responses on some dimensions, so that a complete data set of plausible values are created without any missing values. The aim of this example is to see how well plausible values (including imputed PVs for students with missing responses) recover population correlation parameter. The results are summarised below.

For the multidimensional model, the correlation between plausible values is 0.80, which is also the generating correlation.

For the unidimensional model where missing responses have imputed plausible values, the correlation between plausible values is 0.18. If we only use plausible values for students who have complete data (so there is no imputation), the correlation between plausible values is 0.36. Note that using plausible values from unidimensional models produces worse results than using EAP ability estimates in recovering correlation.

As the unidimensional model does not take information from the other dimension into account, the imputed plausible values for a student with a missing test score is from the estimated *population* marginal distribution. This considerably lowers the correlation. In contrast, in the case of the multidimensional model, the imputed plausible value for a student with a missing test score is from the estimated *conditional* marginal distribution (which has been established with the "correct" correlation parameter between the latent variables), so the plausible values produced reflect the correlation structure of the latent dimensions.

The key message is that if secondary data analysts use plausible values to explore correlations between latent variables, then it is essential that plausible values are produced using a multidimensional IRT model. More generally, for Bayesian IRT models, the specification of the population model must be consistent with the statistics of interest. For example, if we are interested in estimating correlations between dimensions, then a multidimensional model must be used to include the correlation as a parameter in the population model.

Summary

At individual student level, the use of multidimensional item response model does reduce the magnitude of measurement error. But the amount by which measurement error is reduced depends on the test length and the strength of correlation between the dimensions. However, while there is a gain in measurement precision, there is also a bias in EAP ability estimates. If a test is already long (say, more than 50 items), the use of unidimensional item response model may be adequate for the purposes of estimating individual student abilities. Further, it should be noted that, in a multidimensional item response model, the results on any dimension has an impact of the results on other dimensions. Consequently, the estimated ability on one dimension will be closer to the abilities on other dimensions. For some students, this will result in a better estimate (in the sense that it is closer to the true ability). But for other students, this may result in a small bias in estimated abilities. That is, the final ability estimate is no longer just based on what the student did on that test. It incorporates other information as well. This may or may not be desirable, as more explanations will need to be given about how test results are produced. There is also a *perceived* fairness that needs to be considered. For example, if both Student A and Student B received the same test score on reading, but Student B had a higher mathematics score, then Student B's estimated reading ability would be higher if a multidimensional model is used.

At the population level, the cohort mean and variance are estimated equally well whether unidimensional or multidimensional Bayesian item response model is used. However, the correlation between two latent variables is recovered well only with the multidimensional model. When there are missing cases for one dimension and not the other, the multidimensional item response model uses the estimated correlation parameter to draw upon information from the available data in other dimensions for imputing missing scores so that complete data sets without missing values can be produced. Imputed plausible values from a multidimensional model recover the correlation well while plausible values from unidimensional models do not. As a rule, in Bayesian IRT models, the population model for producing student scores for secondary data analysis needs to be consistent with the statistics of interest in the secondary analysis.

Discussion Points

(1) Discuss when multidimensional models should be used in preference to unidimensional models.
(2) Explain why it's possible to have biased estimates but yet smaller RMSE.

Exercises

Q1. Indicate whether you agree or disagree with each of the following statements

Latent correlation refers to the correlation between "true" abilities (i.e., not between estimated abilities)	Agree/disagree
The test reliability for each dimension will be similar whether unidimensional or multidimensional model is used	Agree/disagree
Because multidimensional models draw on information from all dimensions, the estimated correlation from MIRM will likely overestimate the latent correlation	Agree/disagree
Because multidimensional models draw on information from all dimensions, the EAP ability estimates will be influenced by students' scores on all dimensions	Agree/disagree
Multidimensional models produce biased population mean estimates	Agree/disagree
Imputing missing student scores will overestimate the correlations between two dimensions	Agree/disagree

References

Adams RJ (2005) Reliability as a measurement design effect. In: Postlethwaite (ed) Special issue of studies in educational evaluation (SEE) in memory of RM Wolf, vol 31, pp 162–172

Adams RJ, Wilson M, Wang W (1997) The multidimensional random coefficients multinomial logit model. Appl Psychol Meas 21:1–24

Bock RD, Gibbons R, Muraki E (1988) Full-information item factor analysis. Appl Psychol Meas 12:261–280

Embretson SE (1997) Multicomponent response model. In: van der Linden WJ, Hambleton RK (eds) Handbook of modern item response theory. Springer, New York, pp 305–322

Fischer GH (1995) Linear logistic test models for change. In: Fischer GH, Molenaar IW (eds) Rasch models: foundations, recent developments and applications. Springer, New York, pp 131–156

Jöreskog KG (1969) A general approach to confirmatory maximum likelihood factor analysis. Psychometrika 34:183–202

OECD (2012) PISA 2009. Technical report, PISA, OECD Publishing

Reckase MD (2009a) Multidimensional item response theory. Springer, New York

Further Reading

This chapter provides a brief introduction to multidimensional IRT models based on the formulation of MIRM by Adams, Wilson and Wang (1997). This is only one type of MIRM

Reckase (2009) provides a comprehensive discussion on the developments in MIRM more generally, and the topics include formulations of the models, parameter estimations, model fit and equating designs

In addition to MIRM, many analyses for dealing with multiple abilities and multiple cognitive components have been developed. These include Bock and Aitkin's full-information item factor analysis (Bock et al. 1988), Embretson's Multicomponent Response Models (Embretson 1997) and Fischer's Linear Logistic Test Model (LLTM) (Fischer 1995)

Further, confirmatory factor analysis (CFA) (Jöreskog 1969) provides another statistical tool for modelling multidimensional latent trait data

Glossary

Ability This refers to the level of latent trait of a respondent as measured by an instrument for a certain construct. It is usually represented by the total test score in the classical test theory or in terms of logit in item response theory. See Chap. 1.

Assessment framework An assessment framework is a document usually written by subject matter experts and measurement experts. The document typically covers the purpose of the assessment, the target population to be assessed, assessment methods, and most importantly, the definition of the construct to be measured and the content to be covered in the assessment. See Chap. 2.

Balanced incomplete block (BIB) design This refers to a test booklet design where each cluster of items appears in each position of a test booklet, and every pair of clusters appears together in one test booklet. See Chaps. 3 and 13.

Between-item dimensionality For multidimensional IRT models, each item loads on only one dimension of the latent constructs. That is, there is a set of items tapping into dimension 1, and a different set of items tapping into dimension 2, etc. See Chap. 15.

Booklet design This refers to the arrangement of items in test booklets. In particular, in large-scale assessments, curriculum coverage requires many items to be used. In order not to over burden the students to answer a long test, items can be distributed to different booklets and each student is required to take only one booklet. Typically, the items are grouped in blocks or clusters which are then arranged according to a balanced incomplete block design. See Chaps. 3 and 13.

Calibration This refers to the procedure of estimating the item difficulties and the abilities of respondents on a scale of a latent variable. See Chap. 8.

Classical test theory Classical test theory (CTT) refers to the analysis of test results based on test scores. CTT typically includes the notion of the reliability of a test, point-biserial correlation for each item, and test scores for students. See Chap. 5.

M. Wu et al., *Educational Measurement for Applied Researchers*,
DOI 10.1007/978-981-10-3302-5

Cluster sampling Cluster sampling occurs when groups of respondents (e.g., schools or classes) form the sampling units instead of individuals in the population. See Chap. 3.

Codebook A codebook provides information about a data set, such as variable names, variable labels, value coding and what the value coding refers to. It enables any researcher analysing the data to know what the data are and how to access them. See Chap. 4.

Common items One technique for equating tests is to use common items (also known as link items) in multiple tests and then align the calibrations based on these common items. See Chap. 12.

Complex sampling We refer to probability sampling other than simple random sampling as complex sampling. Complex sampling may involve stratifications of the sampling frame, systematic sampling and cluster sampling. See Chap. 4.

Construct (Latent variable) This refers to a trait that cannot be observed directly. The construct of a measuring instrument is what we are trying to measure with the instrument. See Chaps. 1 and 2.

Control script Control scripts are example student responses to extended-response items for the use of marker training sessions. The purpose of using control scripts is to familiarise markers with the marking guide by providing them with guidelines in categorising student responses. See Chap. 4.

Cronbach's alpha Cronbach's alpha is a measure of the internal consistency of a test or a group of items that tap into the same construct. It is one of the most commonly used reliability coefficients in applied studies within the classical test theory. See Chap. 5.

Data cleaning Data cleaning refers to checking for, and rectifying, anomalies in the data. It includes such procedures as value range check, missing values treatment, duplicate record check, inconsistency check and multiple instruments check. See Chap. 4.

Design effect The design effect is the factor by which the sample size of a simple random sample needs to be inflated for complex sampling design in order for the latter to achieve the same accuracy as for a simple random sample. See Chap. 4.

Dichotomous score/data This refers to the response outcomes of respondents to a set of items. The outcomes are classified into two discrete categories (e.g., not present/present, yes/no, and wrong/right). The categories are usually scored as 1 and 0 for the ease of data analysis. See Chap. 7.

Differential item functioning (DIF) An item is said to exhibit DIF when the probability of success on the item differs for two groups of respondents even when the abilities of the two groups of respondents are matched. DIF is caused by different strengths and weaknesses of respondents owing to a number of possible factors, including different curriculum, different personal disposition, experience, culture, language and many other reasons. See Chap. 11.

Embedded-missing items The term embedded-missing items refer to those items being skipped by students while taking a test. See Chap. 4.

Equating When two tests need to be placed on the same ability scale, an equating procedure is required in order to put the parameters of the two tests on the same scale for comparison. See Chap. 12.

Expected a posteriori (EAP) statistic Expected a posteriori (EAP) statistic is a point estimate for a student's ability in the Bayesian IRT approach by taking the mean of each respondent's posterior distribution. This is sometimes used as an ability estimate under the MML estimation method. See Chap. 14.

Expectation Generally speaking, the expectation of a random variable refers to the long run average value under repeated realization of the variable. It is also known as the expected value of the random variable. When applied to the observed scores of student taking a certain test, it refers to the long run average value of the observed scores under repeated administrations of the same test to the same student. See Chap. 5.

Expected score This is the average item score for respondents with a given ability, computed using the theoretical item response function. See Chap. 9.

Facets model This is a class of IRT models that incorporate factors (in addition to item difficulty and student ability) that influence the probability of success on an item. For example, the inclusion of a rater harshness parameter is an example of a facets model. See Chap. 13.

Free calibration This refers to the estimation of item parameters based on the item response data for a test and not linked to any other test results. See Chap. 12.

Generalised partial credit model This is a 2-PL extension of the partial credit model. There are, however, different ways to generalise the partial credit model. See Chap. 10.

Horizontal equating Horizontal equating refers to equating tests aimed for the same target level of students. For example, if a number of tests for grade 4 students are administered, the equating of these tests onto the same scale for comparison is known as horizontal equating. See Chap. 12.

Infit statistics This is a residual-based weighted fit statistics for assessing item fit. See Chap. 8.

Information function Conceptually, this function gives us an idea of how useful an item or a test is for estimating abilities. See Chap. 3.

Item characteristic curve The item characteristic curve (ICC) of an item shows the probabilities of answering an item correctly by respondents across a spectrum of abilities. This curve is often formulated in terms of a logistic function, which looks like an elongated letter S. The ICC is sometimes known as the item response function. See Chaps. 6 and 7.

Item dependency This refers to the violation of the local independence assumption of the Rasch model when the probability of success on an item depends on the response(s) on other item(s). See Chap. 8.

Item difficulty In the dichotomous Rasch model, an item's difficulty is the location on the scale at which the respondents have 0.5 chance of answering the item correctly. The item difficulty is often used to place an item on the scale of the latent variable. See Chaps. 2, 3, 6 and 7.

Item discrimination In classical test theory, item discrimination is a measure of the relationship between the scores on an item and the overall test scores of students. In IRT perspective, it refers to the slope of the item characteristic curve. See Chaps. 5, 7 and 10.

Item fit statistics IRT has an underlying mathematical model to predict the likelihood of the item responses. Statistical tests of fit can be constructed to assess the degree to which responses of an item "fit" the IRT model. Such fit tests provide information on the degree to which individual items are indeed tapping into the latent trait. See Chaps. 2 and 8.

Item invariance This refers to the situation when items are found to perform *in the same way* across different tests. See Chaps. 6 and 12.

Item-person map This is a map that shows the relative positions of item difficulties and the abilities of persons on the same scale. It is usually organised as a map with two panels. The left panel usually displays a distribution of the respondents' abilities, while the right panel displays a distribution of the location of the items. It is also known as a Wright map or variable map in the literature. See Chap. 6.

Item position effect This refers to the situation when an item has different difficulties if it is placed at different positions in a test, say, the beginning and the end of a test. See Chaps. 3, 12 and 13.

Item response theory (IRT) Item response theory assumes an underlying mathematical model to predict the likelihood of the item responses by the respondents according to their abilities and a number of parameters. See Chaps. 2, 6 and 7. Note: We also refer to it as item response modeling

Latent regression The population model in Bayesian IRT specifies that the mean of the ability distribution is formed by a regression-like formula typically containing student background variables. See Chaps. 13 and 14.

Latent trait variable See construct. See Chaps. 1 and 2.

Learning progression When item response data fit the Rasch model, one can write summary statements of skills along the ability scale based on the locations of test items positioned according to their item difficulties. These summary statements are descriptions for a learning progression that typically apply to the population of test takers. It describes the order of difficulty of skills to be mastered and is sometimes known as a proficiency scale in the literature. See Chap. 7.

Level of measurement This refers to how numerical values are assigned to attributes of objects according to some rules. A common treatment is to claim that there are four levels (or scales) of measurement, namely, the nominal, ordinal, interval and ratio levels. The numerical values from different levels of measurement convey different amount of information. See Chaps. 1 and 2.

Linking In this book, linking is used as a synonym with equating. See Chap. 12.

Logit (logit scale) In item response theory, the measurement unit of the scale for ability and item difficulty after the $\log(p/(1 - p))$ transformation is generally known as "logit", a contraction of "log of odds unit." See Chaps. 6 and 7.

Local independence An important assumption for the Rasch model is that the probability of success depends only on a person's ability and an item's difficulty. The probability is not influenced by a person's success or failure on other items, or by factors other than ability and item difficulty. This assumption is generally referred to as the local independence assumption. See Chap. 7.

Mantel Haenzsel test This is a method for detecting differential item functioning. See Chap. 11.

Maximum a posteriori statistics Maximum a posteriori (MAP) statistic is a point estimate for a student's ability in the Bayesian IRT approach by taking the mode of the posterior distribution. See Chap. 14.

Marginal maximum likelihood estimation In some IRT models, there is an assumption of the distribution of the population of abilities. The MML estimation method incorporates this population distribution with the item response function. See Chap. 14.

Marker harshness/leniency This refers to raters' propensities for being harsh or lenient in grading. See Chaps. 3 and 13.

Marking guide (or scoring rubric) This refers to a guideline that is established for scoring purposes. This is usually used in scoring responses to constructed response items, such as short response items or extended essays. See Chap. 4.

Measurement error Measurement error refers to the possible variation in a student's test scores if similar tests are administered. There is always some uncertainty associated with a test score, not because the test contains errors, but because by chance the student may know more or less of the content of a particular test. Measurement errors are typically large for an individual because a test contains limited number of items and hence the possible variation in test scores is usually large. See Chaps. 3 and 5.

Measurement invariance Measurement invariance refers to the invariance of the relative placements of students on the ability scale irrespective of the instruments being administered to them, provided that the instruments all measure the same construct. See Chap. 6.

Multidimensionality When test items tap into multiple constructs, the test is said to be multidimensional. See Chap. 12.

Multidimensional IRT models These are IRT models for measuring multiple constructs (abilities). See Chap. 15.

Not-reached items Not-reached items refer to the missing responses at the end of a test, with the possibility that students ran out of time and never had the opportunity to answer the items at the end of a test. See Chap. 4.

Outfit statistics This is a residual-based unweighted fit statistics for assessing item fit See Chap. 8.

Partial credit model It is a Rasch model formulated to analyze data collected from instruments with polytomously scored items. See Chap. 9.

Plausible values These are random draws from each student's posterior distribution under the Bayesian IRT models. See Chap. 14.

Point biserial correlation This is a classical test theory item statistic assessing the degree to which an item can separate students according to ability levels. See Chap. 5.

Polytomous score An item is said to be polytomous when there are more than two scoring categories. See Chap. 9.

Posterior distribution This is the estimated ability distribution for a student under the Bayesian IRT models. See Chap. 14.

Prior distribution This is the population distribution of abilities. See Chap. 14.

Probability sampling Probability sampling means that every unit (e.g., school/student) in the target population has a chance of being selected, and these chances can be computed according to the sampling design being used. See Chap. 4.

Rasch model This refers to a family of measurement models that have measurement invariance properties. This includes the model for dichotomous data, the partial credit model and the facets model, among others. See Chaps. 6, 7, 9 and 13.

Rating scale model This is in the Rasch models family, formulated to analyze data collected from rating scale instruments. In this book, we regard this model as a special case of the partial credit model. See Chap. 9.

Raw data Raw data refers to the responses given by the respondents to a test instrument before any data processing is carried out. See Chap. 4.

Reliability Reliability refers to the degree to which an instrument can separate respondents by their levels on the construct. See Chaps. 1, 2 and 5.

Response probability In the item-person map, when items are matched to a person to describe the performance of the person on the items, it is usually regarded that the person has a 50% chance of answering those items correctly. This probability is regarded by some as being too low and is changed to a higher value. The probability deemed as appropriate to match a person to the items is sometimes called response probability, or RP in short. See Chap. 7.

Sampling design A sampling design refers to ways the sample of participants is selected from a population for a study. Some examples are simple random sampling design and cluster sampling design. See Chap. 3.

Sampling frame A sampling frame is a document that lists all the units of a target population subjected to sampling. In educational surveys, sampling is usually done by first identifying all schools in which students in the target population are enrolled. The names of these schools, important information (e.g., address, school type, geolocation) and the enrolment size for each grade in each school are then made into a list. This list is known as a school sampling frame. Similarly, a sampling frame of students can be made when sampling students from selected schools. See Chap. 4.

Sampling weight One simple way to understand this is to think of sampling weight as the number of students in the target population represented by a sampled student. See Chap. 4.

Specific objectivity This is one of the properties of the Rasch model which refers to the principle that comparisons between two objects must be free from the conditions under which the comparisons are made. This is sometimes referred to as the invariance property in the literature. See Chap. 7.

Standard error of measurement This gives the degree of uncertainty surrounding a test score or associated with an ability measure. See Chaps. 3 and 5.

Stratified sampling It is a sampling design in which stratification is done by grouping the sampling units (e.g. schools) in the target population into strata, such as by geographical location or by school types (e.g., public, private) to ensure that when samples are selected, each stratum has a representative sample of schools. Sampling is then performed proportionally according to the size of each stratum so as to achieve a more representative sample of the target population. See Chap. 4.

Student participation forms A well documented test administration will include a student participation form that contains students' background information (e.g., date of birth, gender), booklet assignment information as well as test attendance records. The attendance records will be useful in computing the adjusted sampling weights. See Chap. 4.

Sufficient statistics In the context of a Rasch model, this refers to the statistical property that students with the same raw score will be given the same ability estimate in logits, irrespective of which items they answer correctly on the test. See Chap. 7.

Test blueprint The test blueprint is usually a table in which the number (or percentage) of items with respect to various contents of the test is being reported. This can also be done with respect to the cognitive domain that the items belonged. The test blueprint is sometimes known as the two-way specification form when the number of items is reported in a contingency table with respect to both the content and cognitive domains at the same time. See Chap. 2.

Test design Test design refers to the considerations for the number of items in a test, the sample size of students to take the tests, the assignment of tests to students, the arrangement of items in a test and the assignment of markers to test scripts. More generally, the development of the construct, framework and test blueprint are all also part of the test design. See Chaps. 2 and 3.

Testlet A testlet is a set of items that are linked to a common stimulus, usually a common passage, a diagram or a common condition. The presence of testlets within a test often leads to the violation of the local independence assumption under the Rasch model. See Chap. 7.

Two-parameter IRT model This is an IRT model where there are two parameters related to each item: the item difficulty parameter and the item discrimination parameter. See Chap. 10.

Two-stage sampling In educational studies, two-stage sampling refers to the practice where a number of schools are first randomly sampled from target population of schools and then a number of students are randomly sampled from the selected schools. See Chaps. 3 and 4.

Unidimensional test A test is said to be unidimensional if all its items should tap into the same latent variable. This is a required condition for aggregated item scores to be meaningfully interpreted. The scores then reflect an overall performance by the respondent on the whole test. See Chaps. 2 and 7.

Validity Validity is about whether it is valid to use measures of an assessment for the purposes of the assessment. See Chaps. 1 and 2.

Vertical equating This refers to equating tests that are administered between different grade levels, for example, between grades 4 and 5. See Chaps. 12 and 13.

Weighted likelihood estimate of ability Since the maximum likelihood approach to the estimation of ability has been found to be biased outwards, Warm (1989) proposed the weighted likelihood approach as a correction to remove this bias. The corresponding estimate of ability is usually denoted by WLE. See Chaps. 7 and 14.

Within-item dimensionality For multidimensional IRT models, an item may load on multiple dimensions of the latent constructs. See Chap. 15.

The manufacturer's authorised representative in the EU is Springer Nature Customer Service Centre GmbH, Europaplatz 3, 69115 Heidelberg, Germany. If you have any concerns regarding our products, please contact ProductSafety@springernature.com

Printed and bound by CPI Group (UK) Ltd, Croydon, CR0 4YY
30/09/2026
02225446-0009